The ergonomics of workspaces and machines

A design manual

SECOND EDITION

The ergonomics of workspaces and machines

A design manual

SECOND EDITION

E.N. Corlett and T.S. Clark

Taylor & Francis
Publishers since 1798

UK   Taylor & Francis Ltd, 4 John St, London
WC1N 2ET

USA Taylor & Francis Inc., 1900 Frost Road,
Suite 101, Bristol PA19007

British Library Cataloguing in Publication data

A catalogue record for this book is available from the
British Library.

ISBN 0 7484 0320 5

**Library of Congress Cataloging in Publication data
available**

Cover design by Hybert Design and Type

Typeset by Keyboard Services, Luton, Bedfordshire.

Printed in Great Britain by Burgess Science Press, Basingstoke,
on paper which has a specified pH value on final paper
manufacture of not less than 7.5 and is therefore 'acid free'.

Contents

Foreword

The British economy depends on the ingenuity of its workforce. In an era of global financial markets and information technology, capital can be moved swiftly from one part of the world to another. Only labour is fixed.

To be successful in the modern world, Britain must invest heavily in the skills, knowledge and ability of its workers. It must put people first. At work, that means employers can no longer treat their workforce as simply another factor of production. Modern trade unions want their members to be involved in decision-making at work, whether about investment, bargaining over wages, or the design of the workplace.

Our vision of a modern workplace puts people first by fitting jobs to workers, rather than workers to jobs. So trade unions and ergonomists share common interests and common values. We want workplaces, work stations, working tools, and work processes designed with the employee in mind, and we want to encourage our members to get involved in the design process. That is why an understanding of ergonomics is becoming a crucial part of the TUC's approach to better safety standards at work. I hope that designers will begin to take ergonomics on board in everything they do.

This latest edition of the *Design Manual*, with its new emphasis on safety, will contribute to two key goals of the trade union movement – improvement to Britain's economic competitiveness, and better safety standards for our members.

The last thing that the British economy needs is to improve its competitive position by reducing standards. We cannot compete on price or wages with the newly-developing economies, so we should not try to compete by cutting the quality of our products or services, nor by reducing the protection afforded to our workforce. Poor health and safety standards already cost the British economy up to £16 billion a year, according to the Health and Safety Executive. In just one, albeit contentious field, work-related upper limb disorders, or RSI, TUC figures suggest a cost of at least £1.25 billion a year. These burdens on business can be substantially reduced by applying ergonomic principles.

As trade unions we have high hopes of this new guide to the applied ergonomics of practical design. Our members depend on it for their health and their livelihoods.

John Monks
General Secretary
Trades Union Congress

Preface

Every industrial system consists of some or all of the following components: hardware (the physical aspects), software (non-physical aspects), the physical environment and the organization. An objective of the designer is to arrange these components to give a harmonious and efficient operation. An objective of ergonomics is to match, or provide the information to match, the various other parts of the system to the characteristics and abilities of the people involved in it. By utilizing ergonomics the designer's opportunities to create a system which reliably achieves its functions are improved.

This manual is for a wide range of designers but emphasizes data of particular interest to those concerned with workspaces, working equipment and machines. These may be in factories, offices, warehouses, etc. The information is laid out in the sequence in which the designer will most probably require it, starting with a brief introduction, an illustration of the range of data and their interactions with each other, and a checklist for preliminary design decisions. The sections which follow are on workspace design including environment design, design of manual controls, design of displays and information and design for maintainability and for safety. They are cross-referenced and supported by a detailed index and references to other texts.

It should be noted that a very important design requirement is a specification, or brief. The lack of usability of much in the industrial, business and consumer world arises because the designer's instructions have not specified criteria for usability. This manual has been laid out so that the customer for a product or workplace may also specify the ergonomic requirements. In particular, the Ergonomics Check Chart on pp. 9–14 lists briefly the requirements appropriate for such a specification and is designed to assist in the clear specifications of human requirements in a form in which the designer can satisfy them.

The Summary of Contents which follows this preface will provide an overview of the manual and demonstrates its structure and sequence, from which the user may recognize the purposes of each chapter. These chapter summaries are repeated at the start of each chapter, to assist the user in quickly locating any desired point while keeping in mind the overall structure and interactions in the design process.

References, and a bibliography for further reading, are given, but the gradual development of an ergonomics viewpoint and expertise in ergonomics applications will be aided by the introduction into the design office of an ergonomics journal and association with an ergonomics society. The wide applications of the subject, and its contributions to the many aspects of product and workplace, will be better recognized and incorporated by this continuous exposure to its developing applications in business and industry.

Acknowledgements

The authors would like to acknowledge the work of their collaborators in this second edition: Rani Lueder for her contributions on office ergonomics, Stephen Mason, for the chapter on Maintainability, Rob Stammers for his contribution on task analysis and Aileen Sullivan for the chapter on Designing for Safety.

Summary of Contents

This *Summary of Contents* is intended to give the user an overview of the Manual, so that he/she may recognize the range of subjects covered, and the sequence in which they are presented. This sequence approximates to the general sequence of design and the sequence in which the various subjects would be considered. Particular subject areas can also be identified from this summary. To assist the user to retain the structure of the manual, each chapter summary is repeated at the beginning of its chapter. The numbered sub-headings of the summary also represent the chapter sub-headings.

1 Introduction

1.1 Introduction to the ergonomics of machines

1.1.1 Definition

Ergonomics can be defined as the study of human abilities and characteristics which affect the design of equipment, systems and jobs. It is an interdisciplinary activity based on engineering, psychology, anatomy, physiology and organizational studies. Its aims are to improve efficiency, safety and operator well-being.

1.1.2 Content

Every work system consists of some or all of the following components, each of which interacts with the others and with technical, economic and other considerations and constraints.
Ergonomics is concerned with the interfaces and interactions between the human operator(s) and the other components and with the effects of these interactions on system performance.

Figure 1.1 Chart of the interactions in the work system.

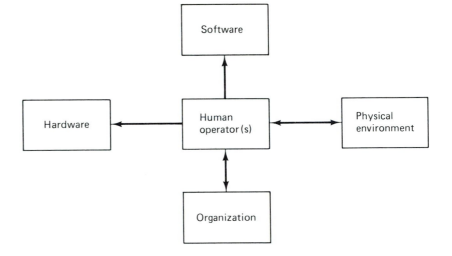

Table 1.1 Components of the work system.

Component	Design areas	Considerations, constraints
Hardware	Design and layout of components	Process, equipment Access
Human Operator(s)	Physical characteristics, abilities, etc.	Body size Strength, work capacity Posture Fatigue and endurance
	Information reception and processing	Senses (vision, audition, etc.) Attention Memory, etc.
	Individual and social characteristics	Age, sex, background, race Skill, training Motivation Job satisfaction and interest Boredom Attitudes, etc.
Software	Error-free performance	Standard operating procedures Instructions Manuals Symbols, etc.
Physical Environment	Safe performance	Temperature Noise Lighting Vibration Atmosphere and ventilation, etc.
Organization	Organization of personnel/production	Work — rest schedules, Pacing, cycle time Shift work Job content Interest Satisfaction Responsibility Social interaction, etc.

1.1.3 Ergonomics considerations in machine design

Figure 1.2 Ergonomics considerations in machine design.

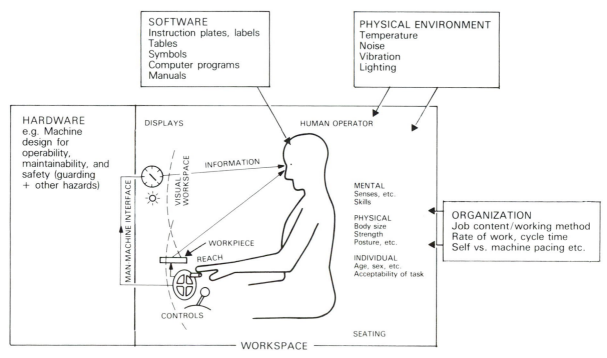

1.2 Main considerations in the application of ergonomics

1.2.1 Important considerations

Good ergonomics design should be regarded as an essential part of good design, not as something separate.

Ergonomics implications should be considered at ALL stages of the design process, especially the early stages.
All design decisions are likely to have some implications for operators or maintenance personnel. It is not sufficient only to consider ergonomics in the final detail stage when all major design decisions have been made.

Ergonomics implications should be considered and discussed by all concerned with or affected by a design.
Co-operation is needed between members of a design team and between designers and the client's production department, personnel, etc. The comments of users are particularly valuable.

Ergonomics requirements should be included in the brief or specification so that their consideration is ensured.
Requirements should be as specific as possible or they may not be adequately considered. If necessary, ergonomics requirements should be budgeted for (e.g., design time, trials, etc.).

There are few simple rules in ergonomics but there are often limits beyond which performance or safety is likely to deteriorate.
Most design decisions involve compromises. If an optimal ergonomics solution is not possible the consequences should be carefully considered, especially if limits are exceeded.

Ergonomics data should be applied intelligently and with caution.
In applying any data care is needed to ensure that the data are applicable to the problem in hand. The origins and assumptions of any data should be examined.

For more difficult problems, where a logical approach alone is insufficient, or where the consequences of error are serious, ergonomics specialists should be consulted.

Mock-ups and prototype trials are important for confirmation.
Using ergonomics information is likely to result in a better first approximation and ultimate design. However, the use of mock-ups, even simple ones, with representative users is valuable for confirming details of fit, reach, layout, etc.

1.2.2 Suggested procedure

Figure 1.3 A suggested iterative approach to design.

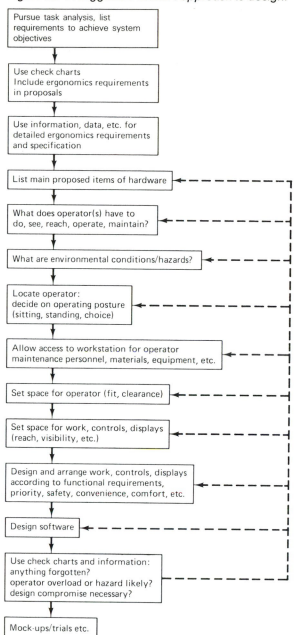

```
Pursue task analysis, list
requirements to achieve system
objectives
        ↓
Use check charts
Include ergonomics requirements
in proposals
        ↓
Use information, data, etc. for
detailed ergonomics requirements
and specification
        ↓
List main proposed items of hardware
        ↓
What does operator(s) have to
do, see, reach, operate, maintain?
        ↓
What are environmental conditions/hazards?
        ↓
Locate operator:
decide on operating posture
(sitting, standing, choice)
        ↓
Allow access to workstation for operator
maintenance personnel, materials, equipment, etc.
        ↓
Set space for operator (fit, clearance)
        ↓
Set space for work, controls, displays
(reach, visibility, etc.)
        ↓
Design and arrange work, controls, displays
according to functional requirements,
priority, safety, convenience, comfort, etc.
        ↓
Design software
        ↓
Use check charts and information:
anything forgotten?
operator overload or hazard likely?
design compromise necessary?
        ↓
Mock-ups/trials etc.
```

1.2.3 Ergonomics specification

Specifies the user population and the levels of human attributes required to be accommodated. Defines the performances required of the users and the equipment, as well as the effects of the design on the people in the environment.

Take each phase of the life cycle of the machine or workspace in turn, e.g. manufacture/assembly; setting up; use, including loading, unloading and servicing; maintenance; dismantling/disposal.

Review proposed design under each of the above headings; assess whether people's capacities are able to achieve the objectives in each phase. Redesign to reduce mis-matches.

Specify users in terms of sex, age, language, skill/training. Set limits or ranges for human and for machine performance. Use Tables 1.1 and 1.2 as a basis for outline, then increase the detail using the appropriate parts of this manual, e.g. maximum noise levels; handling requirements; control/display requirements.

Use Table 1.3 as an aid to priority ranking of specification features. Set out proposals for evaluating how well the specification has been met.

1.2.4 Task analysis

Prior to designing equipment, the designer needs a clear understanding of how people will use it, build it, maintain it or even misuse it. The procedure for defining people's activities with equipment is known as task analysis.

Information from task analysis will be relevant in a number of the areas given in Figure 1.3.

Task analysis seeks to identify all the various sub-tasks required to achieve the system's objectives. The various sources of information required by the users are identified, as are the actions they are required to take, the postures and movements needed and the working conditions under which the work will be done.

The better the quality of this information, the more readily can the designer meet the demands of the specification. The use of a relatively formal procedure is advised, so that the subsequent data can be assessed and verified by users, others in the design team and by the system purchaser. The procedure involves three closely linked stages: information collection, recording and analysis.

Information collection

Be specific about the focus of the analysis, e.g. start-up procedures, operating or maintenance. Choose the most effective information-gathering procedures. Video recording may be ineffective if after hours of analysis it yields little data on the points of concern. Commonly used methods are:

Existing documentation: operating manuals, training manuals, safety reports, previous analyses.

Observation: recording what people do in the specified situations. Study the effects of the various possible circumstances, unlikely as well as likely events.

Questioning: link observation methods with questioning. Ask people to describe what they do, how they do it, what information they use and how they tell if a task has been carried out satisfactorily. Get them to review the results of observations. Questioning can be formal or informal, but should usually be confidential and anonymous, particularly if error or accident information is needed.

Recording: voice, photographic or video techniques are useful, although voice and video can require hours of analysis. But they can be re-assessed by others and repeatedly studied.

The selection of techniques depends on the information needed and the circumstances of the investigation. Field studies are usually the most effective, but simulators, even simple simulations, can be useful. Results from task analysis should be verified by others expert in the field. People being studied should have agreed to take part, should see the results, be fully aware of the purposes of the studies and be assured of personal confidentiality.

Information recording

As information is gathered it may initially be in note form, but eventually it will need to be put into a comprehensive analysis document. This will aid subsequent activities and help communication between those working on the same or related problems.

Figure 1.4 Basic hierarchical diagram, for a task analysis of operating an overhead projector during a lecture. Operations which are not described further are underlined (from Shepherd, 1989).

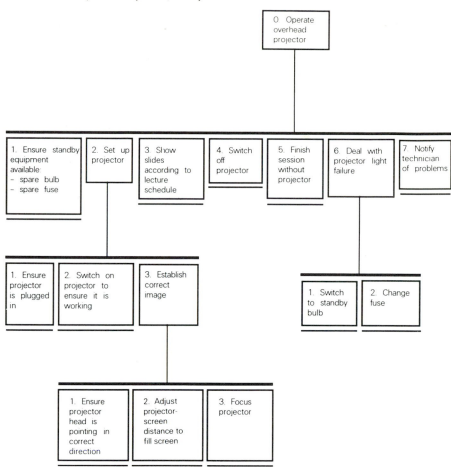

One widely used approach represents tasks in hierarchical diagrams, see Figure 1.4. The aim here is to redescribe the task progressively as more information is collected. The sub-elements of the task are called operations. By going from the general to the specific it is possible to have economy in description, only collecting enough detail for the purpose in hand. The entries in the boxes of the diagram are the names of the operations, and consist of only a few words. Where a task is not to be broken down further, due to its basic simplicity or it not being relevant to the particular study, the box is underlined.

The tasks which are required to operate a projector during a lecture are shown in Figure 1.4, but these are sequenced or selected according to some plan. A plan is a description of how the task's operations are put together as a complete activity, or shows the circumstances under which one operation is performed rather than another.

Plans may be recorded in a cross-referenced table, as in Figure 1.5. They may also be added to the hierarchical table of Figure 1.4, as shown in Figure 1.6. The plan's number is that of the box number to which it refers, the numbers beside each planned step refer to the sub-tasks immediately below in the hierarchy. It can be seen that plans may be stated as simple sequences or as more complex, branching structures.

Figure 1.5 One version of a task analysis table, the plan is in italics (from Shepherd, 1989).

Superordinate	Task analysis – operations–plans	Notes
0	Operate overhead projector *Plan 0:* *At least ½ hr before lecture – 1:* *Immediately prior to lecture – 2:* *As lecture commences – 3:* *If the projector light fails – 4–6–5:* *If other problem occurs that you* * cannot deal with – 4–5:* *At end of lecture – 4 (if on) – 7 (if* * problem) – EXIT.*	
	1. Ensure standby equipment available – spare bulb – spare fuse	Get replacements from the technician
	2. Set up projector	
	3. Show slides according to lecture schedule	
	4. Switch off projector	
	5. Finish session without projector	This should never occur. Unfortunately it sometimes does. Be prepared!
	6. Deal with projector light failure	This is the only fault you should try to deal with yourself
	7. Notify technician of problems	Failure to do this may cause problems for colleagues or yourself later if equipment is unprepared

Figure 1.6 Hierarchical diagram with plans added. The plans, in italics, refer only to those operations that have been broken down into sub-operations.

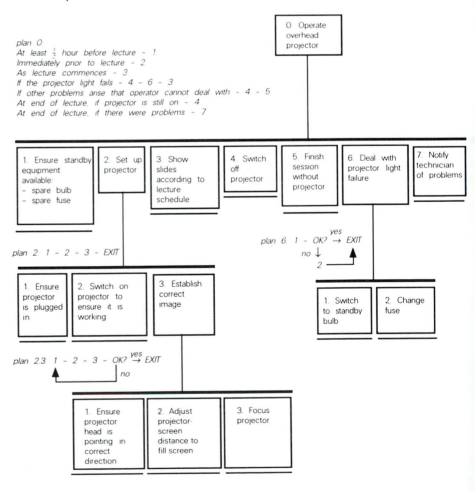

Information analysis

The final phase is where the information is used to yield the basic data for design decisions. For example, if the work requires the design of a set of displays, then the focus will be on information flow across the interface. This is illustrated in Figure 1.7, where extra columns have also been added to the task analysis table to record information inputs to users. These sources may then be used to guide design activity as given in Chapter 5. Another example would be the layout of a set of controls, where the sequence of actions would be important. The sequences illustrated by the task analysis would be isolated and used to guide decisions on optimal layout, for example to minimize the amount of physical movement needed (see Chapter 4). In this case the designer might abstract flow diagrams from the information as a convenient form of documentation. Similarly, design for assembly, for maintenance or for the integration of the new design with existing equipment will all require the contributions from relevant task analyses.

Figure 1.7 Alternative task analysis table showing information flow.

Superordinate	Plan	Operations	Information flow across interface	Information assumed	Task classification	Notes
O. Operate coal preparation plant	1 → 2 and 3 until 4. Do 5 as appropriate and 6 at end of shift	1. Start up plant	→ initiate start ← plant items selected	start-up procedure	procedural	
		2. Run plant normally	← plant operation and monitoring → control information	knowledge of plant flows and operational procedures	operational	
		3. Carry out fault detection and diagnosis	← fault data	some understanding of faults	fault detection fault diagnosis	
		4. Shut down plant	→ initiate shut down	shut-down procedures	procedural	
		5. Operate telephone and tannoy	–	operational knowledge	operational	
		6. Make daily reports	← plant data for log	reporting procedure	procedural	

1.3 Ergonomics check chart

The considerations given in the chart should be satisfied if a design is to be ergonomic.

This chart should be used in preliminary design for preparing the brief or list of general requirements; also it should be used during the design process for checking that nothing has been forgotten.

This chart provides general statements relating to some of the main interactions or considerations indicated in the interaction matrix shown in Table 1.3. Of the design and performance factors listed, functional requirements, safety, operability, size and maintainability are presented in this chart. Other factors can be included as required, depending on the system being designed.

Design and performance factors (and ergonomics factors) are inter-related. The separation of factors in this chart is to simplify analysis and classification. It is not implied that they should be treated separately in design. In particular, safety should be treated as a part of design, not as something separate or as an afterthought.

Table 1.2 Check chart of design and performance factors and ergonomic considerations.

Design and performance factors	Ergonomics considerations	Page
Functional requirements	Assign functions to hardware and operator(s) according to abilities and characteristics of users besides technical, economic and other considerations. Undertake task analysis.	5 6–9
Safety	Utilize safety-by-design procedure.	116
General	Refer to appropriate regulations and standards, etc.	123–125 117
	Identify hazards.	117–118, 118–119
	Make appropriate design decisions.	120–121
	Remove hazard at source if possible	69, 120–121
	or provide barrier (e.g., guard) or separation	120–121
	or provide personal protection.	120–121
	Separate and/or protect from mechanical, electrical, chemical or other hazards.	43–45
	Separate and/or protect from extremes of temperature, noise, vibration and other environmental hazards.	49–60
	Minimize physical, mental and environmental fatigue or stress.	
Physical workspace	Design working position and task to avoid strain and damage to the body, especially the back.	18–37, 108, 111
	Components and materials to be safely handled, manually or mechanically.	41, 104
	Locate hazards beyond longest reach.	43–44
	Openings to be small enough to prevent access to hazards.	44–45
	Provide space for access and emergency exit.	40, 105–107
	Minimize obstruction to physical action and vision.	111
Control design	Assess information requirements.	9
	Choose and design controls for safe and efficient operation, considering requirements for force, speed, accuracy, feedback, etc.	62–70
	Design to avoid accidental operation.	65–68
	Locate to avoid interference.	63–64

Design and performance factors	Ergonomics considerations	Page
	Locate controls for safe, efficient and comfortable operation considering priority, frequency and duration of operation, speed, accuracy and sequence. Locate emergency controls suitably.	63–68
	Controls should move in a direction compatible with display or system movement.	69–71
Display design	Choose, design and locate displays for safe and efficient operation considering operational requirements, type of information presented and what is to be done with the information.	9, 74–77
	Avoid masking (interference) of communication warnings.	33–47
	Provide clear warnings, labels, instructions, manuals, etc.	113

Operability

Body size	Allow range of users to fit work-station and reach work and controls.	10, 26–29
	Note variations with sex, ethnic background, etc.	28
Posture	Avoid fatiguing posture except for infrequent, short duration tasks. Allow changes of posture. Provide support (seating, handle, rails, arm-rests, foot-rests, work-tops, etc.) where possible.	14, 108, 111 18–19, 20, 22, 23–24 19–21 21
Movement	Design for efficient handling (sequence, etc.). Avoid static muscular work. Balance muscle groups (e.g., two-arm operation).	18, 107 22–23, 41 21–22
Strength	Design for variation (e.g., sex differences) and weakest proposed user. Choose limb or muscle group appropriate to the task. Consider maximum vs continuous effort. Consider location, magnitude, direction, distance, frequency and duration of forces. Use power assistance where appropriate.	41–42, 108 41 18, 22–23 63, 65–69

Design and performance factors	Ergonomics considerations	Page
Work capacity/rate	Allow adequate rest pauses or change of task. Machine-paced tasks (work rate determined by machine) to be avoided. Provide buffer storage. Allow for effects of physical environment (temperature, noise, lighting, vibration) on work capacity.	
Visibility	Allow comfortable viewing posture. No visual obstructions: eye positions of users. Objects of suitable size vs viewing distance. Allow for visual defects, spectacles, colour defects.	23–26, 33 23, 33 23, 33 23
Illumination	Provide adequate illumination for task: general background, local, in-built. Design for poorest lighting conditions: brightness, shadow. Provide adequate contrast between object and background: lighting, colour, size, shape. Avoid glare by position and design of lighting, work-surfaces, material. Colours to be appropriate for task, safety, aesthetics.	24–26 57 57 58, 82
Choice and design of controls	Select and design controls according to functional requirements.	62, 63, 65–71
Layout of controls	Arrange work and controls according to priority, functional requirements and comfort: consider the importance for safety, frequency, duration, force, speed, accuracy, sequence and compatibility between controls and displays.	6–9, 63–64, 69–71
Visual displays, information, software	Select and design according to functional requirements, standards, etc.	6–9, 74–77
Layout of visual task and displays	Arrange according to functional requirements for priority, convenience, comfort, importance for attention, frequency, sequence, etc.	77–78
Noise, auditory signals	Auditory signals/displays to gain attention. Auditory environment should not interfere with communication, warnings, etc., or cause annoyance or distraction.	26, 83–84 50–53

Design and performance factors	Ergonomics considerations	Page
Information load	Avoid overloading capacity to receive and process information, e.g., minimize periods of concentrated attention; allow for reduced memory of older operators.	6–9
Size		
Workstation	Choose seated, standing or choice of working position.	5–9, 18–19
	Allow for *range* of users to: fit workstation, reach work and controls, see work and displays.	26, 27–28
	Allow clearances for head, trunk and legs of largest user.	31, 38–40
	Allow reaches for arms and legs of smallest user. Adjustment where appropriate.	31–36, 105
	Allow comfortable vision: viewing posture, viewing angles, viewing distance.	23–24, 37
Access	Space/gangways/aisles/ladders to the workstation, for maintenance and as an emergency escape.	38–40, 104
Equipment	Portable equipment, tools and protective equipment to be appropriate for users, tasks and space.	41, 59–60
	Controls, displays, seating, etc. to be of recommended sizes.	65–69, 78–99
Components	Convenient size and weight for handling in installation, operation or maintenance. Modular construction where possible. Mechanical handling where appropriate. Objects to be of suitable size for viewing in worst conditions or use viewing aids.	41–42, 111–112
Maintainability	Specify design criteria for maintenance; incorporate design for maintainability into main design procedure; identify critical maintenance operations.	101–113
Access	Provide access to all parts for maintenance. Allow priority of access according to design life, probability and consequences of failure. Provide warnings of failure. Consider location of repair site, workshop, factory.	38–40, 104–112

Design and performance factors	Ergonomics considerations	Page
Space	Provide space for maintenance task for: maintenance personnel, operating tools, removing components and openings. Provide access to point of repair: walking, climbing, crawling.	19, 22–23, 38–40 111
Posture	Working posture to be appropriate for nature and duration of task: standing, sitting, kneeling, lying. Avoid interference with other operators.	18, 21–22, 105
Lifting and handling	Components to be suitable size and weight for manual handling where appropriate. Provide lifting gear and lifting eyes if necessary.	41, 104, 111–112
	Covers, cases, fastenings and connectors to be easily removable and replaceable.	104–105, 100, 111, 112
Instructions and manuals	Provide instructions, labels and manuals for safe and effective maintenance.	112–113
Physical environment	Allow for environmental conditions and safety of maintenance task – provide protection.	50–60, 102–103, 104

1.4 Interaction matrix

The matrix below (Table 1.3) shows the main interactions between ergonomics factors and design and operational factors relating to machines. Other factors (e.g., organizational) may be included, depending on the system.

A general interaction between the ergonomics factors may be assumed. The strength of the individual interactions will depend on the system.

Table 1.3 *The main interactions between ergonomics factors and the design and performance factors.*

Ergonomics factors	Design and performance factors										
	Functional require-ments	Cost	Size[a]	Operability	Safety	Maintain-ability	Reliability	Manu-factur-ability	Quality control	Market-ability, accept-ability	Aesthetics
General human characteristics											
Human vs. machine operation	●	●	○	●	●	●	●	●	●	●	
User population characteristics (age, sex, background, toleration)	○	○	●	○	●	●			●	●	○
Skills, selection, training	●	●		●	●	●		●	●	●	
Physical Workspace											
Body-size variation	○	○	●	●	●	●				●	
Reach, clearance and fit	○	○	●	●	●	●				●	
Postural comfort	○		●	●	○	○				○	
Seating design	○		○	○	○					○	
Strength: limits and variations	●			●	●	●				○	
Physical work capacity, endurance	●	○		●	●	○	○			○	
Control design	○	○	○	●	●	○	○				○
Layout of work and controls	○	○	○	●	●	●	○		○	○	○
Visual workspace, display and Information											
Visual abilities and defects	○	○	○	●	●	○			●		
Visual task design	○	○	○	●	●	○	○	○	○	○	
Visual display design	○	○	○	●	●	○	○	○	○	○	
Layout of visual tasks and displays	○	○	●	●	○	○		○			○
Control – display compatibility	○	○		●	●		○				
Passive displays (labels, symbols, instructions, manuals)	○	○		●	●	●	○				○
Auditory signals and displays (attention, processing, memory, etc.)	○	○		○	●	○	○				
Information load	●			●	●		○				
Physical Environment											
Lighting (recommended illumination, contrast, colour, glare)	○	○	○	●	●	●	○		●	○	○
Temperature (dry bulb high/low, radiant high/low, humidity, air speed)	○	○	○	●	●	●	○		○	●	
Noise (dangerous levels, masking of signals)	○	○	○	●	●	●	○		○	●	
Vibration (damaging effects, interference)	○	○	○	●	●	●	○		○	●	
Organizational											
Layout and flow of personnel, material and plant	○	●	●	●	○	○					
Rate of work (pacing, buffer stocks, shifts)	○	●		●	○	○			○	●	
Job content	○	●		○	○	○	○			○	
Inspection system/Quality system	○	●					●		●		

○ = interaction.　　● = strong interaction.

[a]Size refers to the size of the structure, access, workstation or components.

2 Workspace Design

2.1 Design decisions and principles

2.1.1 General arrangements

The worker should be able to maintain an upright and forward-facing posture.

Avoid unbalanced postures (leaning or twisting), and the need for muscle activity to support the legs or upper arms. Small and/or precise movements require support of the limb(s) involved.

Where vision is a requirement of the task, the necessary work points must be adequately visible with the head and trunk upright or with the head inclined slightly forward.

All work activities should permit the worker to adopt several different, but equally healthy and safe, postures without reducing capability to do the work.

Work should be arranged so that it may be done, at the worker's choice, in either a seated or a standing position. When seated, the worker should be able to use the back-rest of the chair at will, without necessitating a change of working movements.

The weight of the body, when standing, should be carried equally on both feet, and foot pedals should be designed accordingly.

Work should not be performed consistently at or above the level of the heart; even the occasional performance where force is exerted above the heart level should be avoided. Where light hand work must be performed above heart level, rests for the upper arms are a requirement.

Work activities should be performed with the joints at about the mid-point of their range of movement. This applies particularly to the head, trunk and upper limbs.

Where muscular force has to be exerted it should be by the largest appropriate muscle groups available and in a direction co-linear with the limbs concerned.

Where a force has to be exerted repeatedly, it should be possible to exert it with either of the arms, or either of the legs, without adjustment to the equipment.

2.1.2 Working position

Design a work situation based on the following criteria.

Operator's choice

It is preferable to arrange for both sitting and standing.

Sitting

Where a stable body is needed:
for accurate control, fine manipulation;
for light manual work (continuous);
for close visual work – with prolonged attention;
for limited headroom, low work heights.
Where foot controls are necessary (unless of infrequent or short duration).
Where a large proportion of the working day requires standing.

Standing

For heavy, bulky loads.
Where there are frequent moves from the workplace.
Where there is no knee room under the equipment.
Where there is limited front – rear space.
Where there is a large number of controls and displays.
Where a large proportion of the working day requires sitting.

Support seat

Where there is no room for a normal seat but a support is desirable.

Computer use

When both a computer and paper-work are required, a corner work-surface can allow flexibility of viewing distance for the screen and provide ready access to surfaces for the documents or other paper usage.
Where a computer is placed on a desk, to avoid a twisted posture, for example by the user putting keyboard or screen to one side of the desk, provide room for documents beside the keyboard, and depth to get the screen far enough back.
Exploit angled workstations and swivelling chairs, as well as retracting keyboard supports.

19

2.1.3 Access

For the user, an adjustable height work-surface is more effective than an adjustable height seat, providing that the seat fulfils the requirements of Section 2.1.5.

To workstation
To workpoints
For maintenance

In general, allow space for the largest user(s) and make allowances for equipment, tools, temporary storage of components, etc. Access should be possible from the chosen work position while maintaining an upright posture and without excessive arm reach.

Height.
Shoulder width.
Trunk girth.
Hand dimensions.
Arm dimensions.
Aisle, gangway,
   ladder dimensions.
Size of openings, etc.

Round off edges which may be used for support or impacted by users. For desks and consoles, round off vertical surfaces if users' legs may impact when sitting.

Accommodate individual differences; overweight people may need special seating, for example deeper seat depth and higher lumbar support adjustment, or extra space. Short people may also need special seating as low as 400 mm, and lower desks. Allow for handedness. Allow for wheelchair access or space to store disability aids.

For maintenance allow room for activity plus the largest component which may have to be moved, plus handling equipment. See Chapter 6, *Maintainability*. Use appropriate anthropometric data.

2.1.4 Clearance

Vertical

The minimum space between the floor and an overhead obstruction must allow for the largest user plus footwear and headgear (NB roof heights have architectural requirements).

Head height (standing).
Seat/head height (sitting).
Knee clearance (sitting).

Lateral

Design for the largest user plus an allowance for movement and equipment.

Hip width, shoulder width.

Forward

Design for the largest user plus an allowance for movement and equipment. Provide a recess for the foot.

Forward trunk dimension.
Seat-back – worktop dimension.
Foot length.

Hazards	The hazard must be beyond the reach of the longest arm (free reach). The opening size and distance to the hazard from the guards must be such that the hazard cannot be reached by the longest, smallest diameter finger.	Arm length. Shoulder length. Hand, finger dimensions. See Section 2.4.5.
Knee-well width	Design for the seat width plus movement.	Seat width, hip width.
Knee-well depth	Design for the longest thigh plus stretching the lower leg.	Thigh length, leg length, foot length.
Seat–worktop vertical clearance	Allow for the largest thigh thickness and for raising the knees, if the pedals are high, and crossing legs.	Thigh thickness.

2.1.5 Seating

Adjustability	The seat should be adjustable both vertically and horizontally in relation to the work.	
Seat width	Design for the largest hip.	Hip width.
Seat length	Design for the shortest thigh.	Thigh length.
Seat height	If work-surface is adjustable, design with sufficient vertical adjustment for the 95 %ile range to sit so that their feet are flat on the floor, on pedals etc. If a fixed height, provide adjustable seat to achieve appropriate elbow height in relation to work-surface, (see Section 2.1.6).	
	Provide large foot-rests, adjustable to allow 95%ile range to sit with feet flat on foot-rest. Set pedals within same range. Provide space in front of and under seat for legs to stretch.	Lower leg length.
Back-rest	Adjustable support for the lower back is essential. Support pad should contact the user's lumbar curve.	Seat to lumbar curve.
	If horizontal forward thrust is required, a high back-rest should be considered.	
Arm-rest	Provide a padded arm-rest. Avoid the necessity for support only via the point of the elbow.	Elbow height.

2.1.6 Work height

The height and angle of work-surface depend on the task. If standing is essential for part of the task, design for high seated access also, i.e. knee clearance under surface and foot support in particular.

Sitting

If work surface is fixed, design for above average seated eye height, provide adjustable seat and foot-rest for shorter people.
If fine work requires close vision or magnifier, provide arm-rests to avoid holding up unsupported arms.
Tilt surface forward where possible to reduce forward bend of trunk, raised upper arms and bent wrists.

Standing

Set work height for appropriate viewing distance and work requirements, see Sections 2.1.8 and 2.4.2.
Use eye height above average, then provide platform for shorter people.

Waist height (height of highest obstruction to forward bending).
Shoulder height.
Eye height.
Arm length.

2.1.7 Reach

Arm (one or two)

Design for the shortest arm and according to postural considerations and task requirement.
Guards should not cause trunk deflection.
In extreme forward reach there should be clearance for the head.
Guard opening width should not require sideways leaning.

Shoulder – grip or finger tip.
Upper arm, forearm, hand, finger lengths.
Shoulder/elbow joint rotation.
Trunk bending/twisting.

Shoulder width.

Leg

When seated, set reach to pedals for shortest lower leg length; for taller people adjustable seat, and adjustable work-surface to pedal distance, are required.
Provide horizontal pedal adjustment to allow for variations in buttock to knee length.

Arm reach zones	For placing work/controls of highest priority, highest frequency, longest duration, large force, high speed and accuracy: select the most comfortable reach zone for sitting or standing upright, facing forward, with the forearm below the heart and not greater than 45° to the side, and the elbow angle at about the midpoint of the range.	Sitting/standing.
	For low priority, low frequency, short duration activities select the most comfortable reach zone when at the limits of reach – the trunk bent and/or twisted.	Trunk bending/twisting.

2.1.8 Visual workspace

Eye position	Usual viewing range is 350 mm to 500 mm; closer for detail, further away for large objects.	
Standing	Set the eye position for tallest, shortest and average user according to the body position chosen above.	Eye height, standing.
Sitting	Allow for the forward adjustment of the seat.	Eye height, sitting.
Obstructions	Check for obstructions to the line of vision between the object and the eye for a range of users.	

Viewing posture

Zones of vision	For work and displays of high priority, frequency, long duration, high speed and accuracy, etc., select the most relaxed viewing position when the head is upright, facing forward and slightly inclined.	Cones (angles) of vision.
	The normal line of sight should be 10° below horizontal for standing, 15° for sitting. This zone is best for attention, scanning, ability to see detail, colour (foveal vision, i.e., centre of eye used), and distance viewing.	
	Range for angle of line of sight: between horizontal and 40° below is preferred, down to 60° below for short periods. Depends on lighting, location of glare sources, the task and the user.	
	Medium priority visual zones should involve small eye/head rotation suitable for occasional reference.	Eye rotation.

	For low priority, low frequency, short duration viewing, etc., low priority visual zones involving head and/or trunk rotation/elevation may be used.	Head/trunk rotation.
Viewing distance	This must be appropriate for the size of the object and the illumination, etc. It must not be closer than the furthest near point (recedes with age) of the user range. Use optical aids where appropriate. For computer screens, viewing distance is approximately 500 mm; bifocal users may need the screen closer and lower with a backward screen angle. Consider special spectacles.	Near point. Accommodation.

Object

Size	The size must be appropriate for the viewing distance and conditions.	Acuity (angle subtended at eye – ability to see detail).
Contrast	It must be discriminable against the background in the poorest conditions by colour, luminance (brightness, contrast), shadow, shape, texture or reflectance.	
Movement	This may be directional or vibrational, and leads to a drop in acuity. Brief pauses can reduce interference with vision	Speed of movement. Direction of movement: better towards the viewer than across the field of vision.
Visual defects	Allow for colour deficiencies (mainly red) in designing displays, etc. Allow for reduced accommodation (hardening of the eye lens) with age. Allow for spectacles (especially bifocals).	

2.1.9 Illumination

General *Localized* *In-built*	Provide adequate illumination according to the requirements of the task (priority, duration, safety, etc.).	Illumination.
Acuity	The ability of the eye to detect detail increases with illumination.	
Contrast	The ability to discriminate between an object and its background is enhanced by increased contrast, not just by increased illumination.	

Direct glare	Avoid by altering the position of the lamp relative to the eye (the source should not be close to the line of sight) *or* by increasing the size of the source *or* by choosing diffused lighting *or* by using screens.	Glare indices recommended in CIBS, 1984.
Colour	The colour spectrum of the lamp must be appropriate for the colour requirements of the task and the colours of the components.	
Offices	Eye discomfort is a major complaint, particularly in computerized offices. Lighting selection is influenced by the individual user, kind of work done, available ambient lighting.	See Section 3.4.3, *The main factors in the visual environment.*
Computer users	Avoid glare from bright light sources, including windows, in front of or behind the user; the latter will introduce reflections, as will light-coloured clothing. Older users are often more sensitive to glare.	
	With age comes a need for more illumination. Provide adjustable light levels, particularly for local lighting.	
	Keep range of luminances within 10:3:1. The brighter the computer screen, the brighter can the work area be.	See Section 3.4.3, *Luminance.*
	Provide non-reflective or matt finish surfaces.	
	Recommended maximum level for downlighting where computer screens are used is 500 lux. With reversed polarity screens and intensive paper usage, up to 750 lux may be acceptable.	
	For screen-intensive tasks with little paper usage, use lower ambient levels.	
	For detail, e.g. multimedia or CAD work, use lower ambient levels. When uplighting is used, ambient levels can be lower as there is more perceived brightness than with downlighting.	
	If documents are of poor visual quality (e.g. small print, coloured paper), then higher local light levels are needed. Local (task) lighting should be directed rather than diffused, but not focus on a small area. The source should not be in the user's line of sight.	

WORKSPACE DESIGN

25

The intensity should be controllable in some manner and give good colour rendition.
The fitting should not occupy the user's preferred work area.

2.1.10 Noise in offices

Objectives of design are to promote speech audibility and privacy, reduce distractions and stress, and prevent hearing loss. Sources of interference and ambient noise are office machines, speech and telephones, local traffic and heating and ventilating equipment.

See Section 3.2, *The auditory environment*.

For telephone use, ambient noise should be below 47 dB(A). For open plan offices up to 56 dB(A) is feasible.

See Section 3.2 for symbols and measurements.

Noise reduction: enclose noisy equipment; use rubber mats to dampen noise transmission.
Use sound absorbent finishes on walls, partitions and ceilings to reduce sound reflections and reverberation.
In open plan offices use dividing panels at least 1.5 m high and 2.4 m in length, to provide 'sound shadows'.
Orient workstations so that users are about 5 m apart, do not face each other, and do not face exterior windows if traffic noise exists.
A very quiet office can allow interference between users. Masking (background) noise is desirable at levels below interference levels, e.g. around 40 dB(A).

2.2 Body-size variations

2.2.1 General

Design for a *range* of users, not just the average (e.g., designing for average reach means that the half of the user population with shorter arms cannot reach).
Use the data available as a basis for judgement and first approximation. Confirm design details with adjustable mock-ups and representative users where possible.
For other than UK population, use appropriate population anthropometric dimensions; check universality, e.g. not just young males from military database, and construct overlays as in Section 2.3.3. For CAD database or manikins, check source, universality and date of survey. Population dimensional changes can be observed in databases as little as ten years apart.

To estimate dimensions of a population for which only the mean and standard deviation of the stature are given, calculate E1 and E2 as shown below, and multiply by the mean and standard deviation of the unknown population's stature respectively.

$$E1 = \frac{\text{Mean of the required known dimension}}{\text{Known mean stature}}$$

$$E2 = \frac{\text{Standard deviation of required known dimension}}{\text{Known standard deviation}}$$

This ratio-scaling technique is due to Pheasant (1986), see *Bibliography*.

2.2.2 Variation

General

Body size varies with:
Body type – endomorph (obese), mesomorph (muscular), ectomorph (thin);
Age – child, youth, adult, elderly;
Sex – male or female;
Nationality – e.g., British, Swedish;
Ethnic origin – e.g., European, Oriental;
Occupation/social group – e.g., general civilian, military, industrial, clerical, etc.
Other factors – e.g., diet, deformities, disabilities.

Percentiles

It is uneconomical and impractical to design for extremes (large or small). It is common to design so that 90% (or 95%) of the proposed user population can be accommodated, i.e., the smallest and largest 5% (or 2·5%) are excluded. The smallest acceptable size is then the 5th (or 2·5th) percentile and the largest acceptable size the 95th (or 97·5th) percentile.

Correlations

For crude purposes the lengths of body segments are generally taken to be proportional to height and the segment sizes proportional to body weight. In practice there is variation in body segment proportions. This may be allowed for in design by considering, e.g., the imaginary largest person with the shortest arms.

2.2.3 Body links

The main links for which dimensions are required are listed in the body links diagram, Figure 2.1. Other dimensions may be required, e.g., hand, for designing tools, handles, etc.
For practical purposes the body can be approximated to a set of links and joints. Muscles act across the joints and exert torques to maintain posture, move the links and exert forces. Limitations of this approach, e.g., a non-constant centre of rotation or a flexible spine can be allowed for in practical design situations.

Figure 2.1 *The main body links for consideration in ergonomic design.*

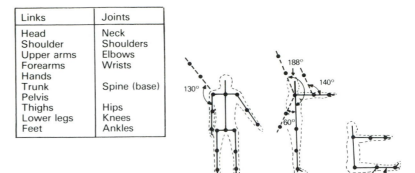

Links	Joints
Head	Neck
Shoulder	Shoulders
Upper arms	Elbows
Forearms	Wrists
Hands	
Trunk	Spine (base)
Pelvis	
Thighs	Hips
Lower legs	Knees
Feet	Ankles

Typical joint rotation limits are shown (some variation between individuals). (See Van Cott and Kinkade, 1972; and NIOSH, 1981.)

Postural comfort   Greatest in the resting position or with joints at the midpoint of the range.

Postural discomfort   At extremes of joint rotation.
When limbs are furthest away from the trunk (especially above the head).
(Consider in terms of moments about joints and in terms of the muscles providing the resisting force.)
Static work, especially if the posture is uncomfortable or a change of position is not possible or a large force of long duration is exerted.

2.2.4 Limitations of data

General   Body-size data cannot be used alone without consideration of the nature of the activities being carried out – frequency, duration, postural comfort, rest pauses, etc. (see the *Ergonomics check chart*, p. 10).

Availability   Data are often unavailable for particular user populations, e.g., various occupational, social or ethnic groups.
Data from similar groups can be used as a basis for judgement but caution is needed and trials are advisable.
Some data are difficult to obtain without a literature search or may be unpublished or confidential in-house work.

Origins   Caution is needed in using design recommendations or data unless their origins are thoroughly explained – they may not be for a compatible situation. Use a variety of sources if in doubt.
Often single values are recommended without tolerance limits.
The surveys may have been carried out on too small a sample.
Many data have been obtained for military (particularly US) populations from which the smaller and obese will have been excluded. Care is needed in applying such data to general populations.

Measurements   For standardization, measurements are frequently of dimensions of nude subjects in an upright posture.
Data are mostly of static dimensions rather than dynamic (reach, etc.) measurements.
Data in various surveys may not be directly comparable, e.g., different dimensions, reference points, origins are used.
Allowances are needed for task requirements, slumped posture, dynamic reach, postural comfort and change of posture, clothing, etc.
High accuracy is not possible owing to the variability of body sizes, arbitrariness of percentile cut-off points, variation of height during the day due to spinal compression, slump causing up to 50 mm change in eye height, variation in the toleration of discomfort, the need to compromise when making design decisions, etc.

Applications

Some translation of data is needed in order to apply them on the drawingboard.

When using human data presented as part of a CAD package, take care to identify the source and adapt them if necessary to the target population. Also note that few packages differentiate between feasible and infeasible, acceptable or uncomfortable postures. What is on the screen is an assumption about human behaviour, subsequent early testing is advisable. No current CAD human body representation contains controls to ensure that the postures shown are what a person would adopt.

2.2.5 Conclusions

Design for the appropriate population.

Use data, if available, or make judgements from data for similar groups.

Carry out trials with representative users if possible.

See the data references in the *Bibliography*.

2.3 Suggested methods of workspace design

2.3.1 Introduction

This procedure, together with *Design decisions and principles* (p. 18), is offered to assist in setting out the workspace on the drawingboard or CAD screen.

A simple approach from first principles is presented, which may be used for applying data obtained from references, surveys, trials, etc.

A direct check method using overlays is also presented.

These can be programmed for a CAD system, or can be drawn to scale for superimposition on a screen or on a printed diagram.

The aim should be, within the limitations described, to allow a comfortable and efficient working posture and method for the proposed range of users.

The use of data should at least produce a good first approximation which may be confirmed by trials.

2.3.2 Design procedure

Given ranges of body lengths and sizes, the limits of rotation about joints and design principles, workspace limits (clearance, reach, visibility, etc.) can be drawn out.

Use overlays and references for dimensions.

Use the following reference points for defining the workspace for large, small and average users:

    Floor.

    Seat (intersection of lines of seat and seat back).

    Eye.

    Shoulder (centre of rotation for reach).

Figure 2.2 Flow chart of design procedure.

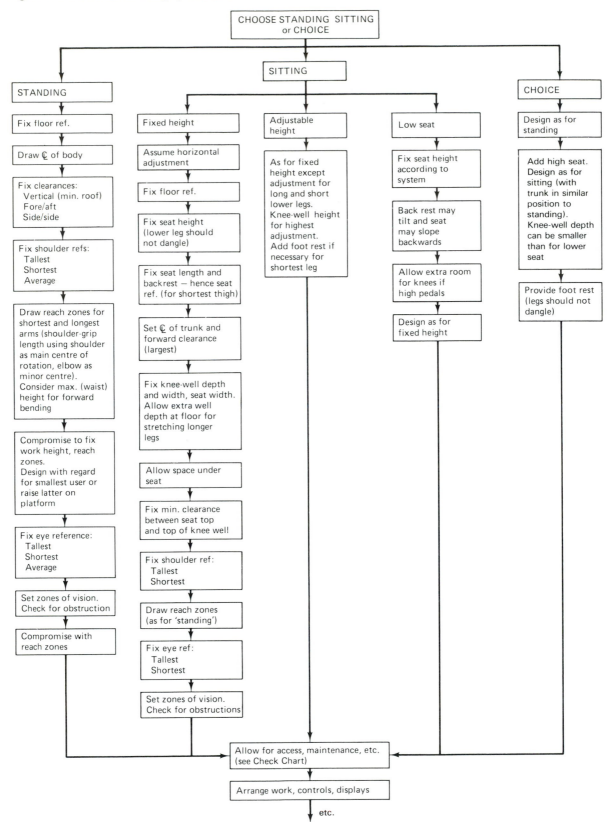

WORKSPACE DESIGN

CHOOSE STANDING SITTING or CHOICE

SITTING

STANDING

Fix floor ref.

Draw ₵ of body

Fix clearances:
Vertical (min. roof)
Fore/aft
Side/side

Fix shoulder refs:
Tallest
Shortest
Average

Draw reach zones for shortest and longest arms (shoulder-grip length using shoulder as main centre of rotation, elbow as minor centre).
Consider max. (waist) height for forward bending

Compromise to fix work height, reach zones.
Design with regard for smallest user or raise latter on platform

Fix eye reference:
Tallest
Shortest
Average

Set zones of vision. Check for obstruction

Compromise with reach zones

Fixed height

Assume horizontal adjustment

Fix floor ref.

Fix seat height (lower leg should not dangle)

Fix seat length and backrest — hence seat ref. (for shortest thigh)

Set ₵ of trunk and forward clearance (largest)

Fix knee-well depth and width, seat width. Allow extra well depth at floor for stretching longer legs

Allow space under seat

Fix min. clearance between seat top and top of knee well!

Fix shoulder ref:
Tallest
Shortest

Draw reach zones (as for 'standing')

Fix eye ref:
Tallest
Shortest

Set zones of vision. Check for obstructions

Adjustable height

As for fixed height except adjustment for long and short lower legs.
Knee-well height for highest adjustment.
Add foot rest if necessary for shortest leg

Low seat

Fix seat height according to system

Back rest may tilt and seat may slope backwards

Allow extra room for knees if high pedals

Design as for fixed height

CHOICE

Design as for standing

Add high seat. Design as for sitting (with trunk in similar position to standing). Knee-well depth can be smaller than for lower seat

Provide foot rest (legs should not dangle)

Allow for access, maintenance, etc. (see Check Chart)

Arrange work, controls, displays

etc.

(see *Introduction, Suggested procedure,* p. 5)

2.3.3 Direct check methods (overlays)

General

The design principles and procedures, with suitable body-size data, can be used for predictive design from first principles.

For quick and direct checks or approximations the workspace drawings may be used indirectly, as transparent overlays. They can also be put under tracing paper as a guide during layout.

These drawings are based on the design principles and procedures and generalized data for UK adults. Judgement has been exercised in their construction and, for reasons already stated, a high degree of accuracy is not possible. However, for practical design situations, where compromises are inevitable, they present a useful first approximation. The drawing can easily be modified by designers for particular situations or user populations and, by using the grid as a guide, can easily be reproduced to any scale.

Each drawing consists of a plan, side and rear sectional elevation set out on a 200 mm grid using floor, seat, eye and shoulder reference points (see Section 2.3.2 *Design procedure*). High, medium and low priority reach and vision zones are drawn according to the *Design decisions and principles* (p. 18).

Reach and clearance

Reach zones are drawn for one arm (for two arms reduce the reach to the side).

Reach is drawn to the centre of the grip (for finger operation extension is possible).

Reach and clearance zones are presented for:

    Standing (fixed floor):
      (a) General UK population: small female (2·5 %ile) – large male (97·5 %ile).
      (b) Larger UK population: small male (2·5 %ile) – large male (97·5 %ile), or average female – large female.
    Sitting: For both population groups as for standing. For fixed, adjustable or high work chair.

Alternative reach zones and clearances may be drawn in proportion to arm length, body size, etc.

Vision

The zones of vision can be used for sitting or standing although the normal (relaxed) line of sight is 10° below horizontal for standing and 15° for sitting.

Use visual overlays in conjunction with reach overlays using eye reference point and check for compromises, obstructions, etc.

Preparation of overlays

Prepare a grid of appropriate size for the scale of drawing:

Scale	Grid size (mm)
1/2	100
1/4	50
1/5	40
1/10	20

Copy the drawings provided using the grid as a guide or produce modified versions for other user populations;
or use reduction copying or other methods of preparation.

Reference points

Reference Points	Eye	Floor	Seat	Shoulder
General	ERP	FRP	SRP	SHRP
Small (2.5%) female	E2.5F	F2.5F	S2.5F	SH2.5F
Small (2.5%) male	E2.5M	F2.5M	S2.5M	SH2.5M
Large (97.5%) male	E97.5M	F97.5M	S97.5M	SH97.5M

Some reference points are not shown on the drawings but may have been used in their construction.

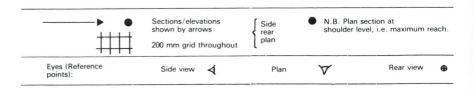

Table 2.1 *Zones of reach and vision according to frequency of use.*

	Reach zone	Visual zone
High	For locating work and controls of highest priority, highest frequency and duration of operation, large force *or* high speed *or* high accuracy. Most comfortable reach zone, sitting or standing upright facing forward forearm below heart and < 45° to side, elbow at mid-point range or slightly bent.	For locating work and displays of highest priority, frequency, duration, etc. Most relaxed viewing posture: head upright or slightly inclined, facing forward. Best zone for scanning, attention, acuity, colour vision and distance viewing.
Medium	For locating work and controls of medium priority and frequency, duration, etc. Sitting/standing upright. Small eye/head rotation. Arm at limit of reach to front and sides. Forearm allowed above shoulder.	For locating work and displays of medium priority, duration, etc. Zone viewed by small eye/head rotation.
Low	For locating work and controls of low priority, low frequency, short duration, low force, speed and accuracy. Maximum reach by bending and/or rotating back.	For work and displays of low priority, frequency, duration, etc. Zone viewed by rotation/elevation head and trunk rotation.

Figure 2.3 Reach and vision zones.

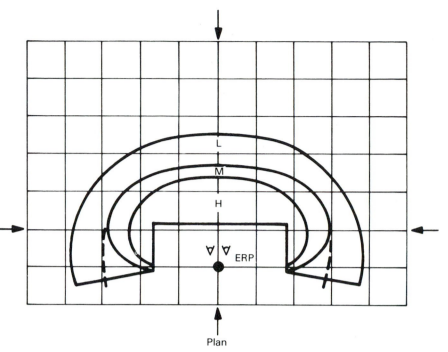

Plan

WORKSPACE DESIGN

(a) STANDING REACH ZONES suitable for general UK adults: small (2·5%ile) female — large (97·5%ile) male.

0·2 m grid

The H zone can be increased vertically if the smallest user is raised on a floor platform.

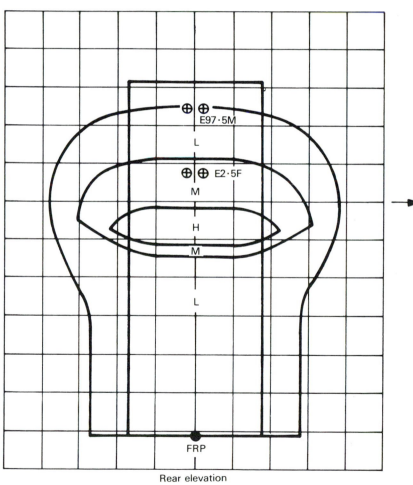

Rear elevation

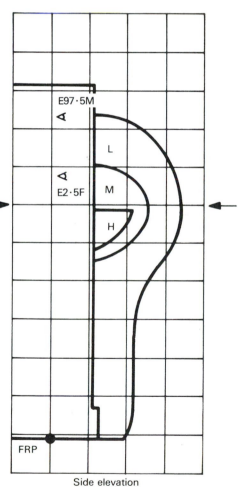

Side elevation

33

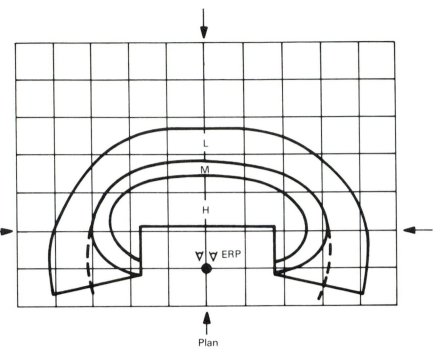

Plan

(b) STANDING REACH ZONES
suitable for general UK adult
males or average — large
females:
small (2·5%ile) males —
large (97.5%ile) males.

0·2 m grid.

The H zone can be increased vertically
if the smallest user is raised
on a floor platform.

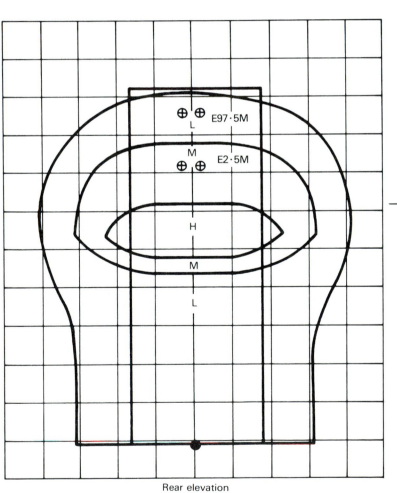

Rear elevation

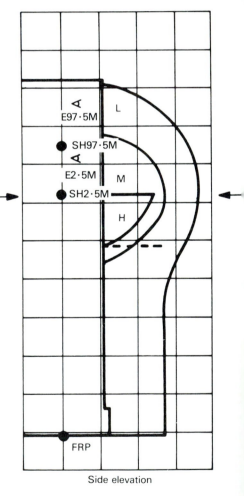

Side elevation

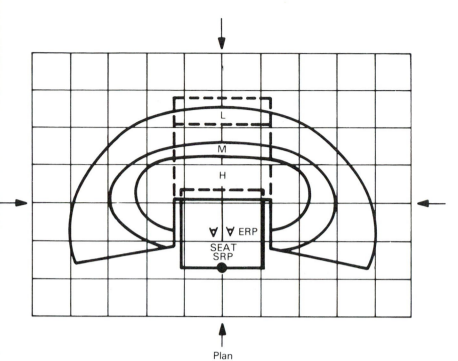

Plan

(c) SITTING REACH ZONES
suitable for a general work chair
(fixed or adjustable)
or high chair:
general UK adults (M or F).

0·2 m grid.

*See Key, notes on use.

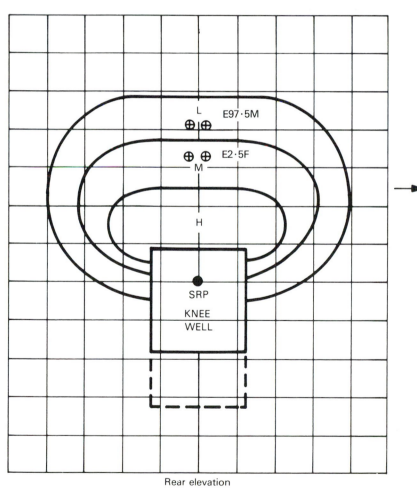

Rear elevation

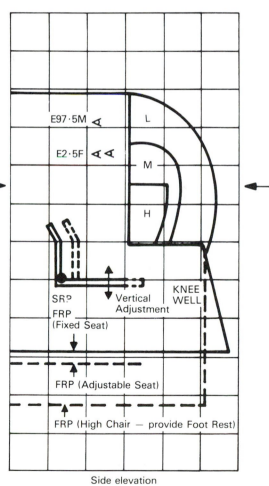

Side elevation

35

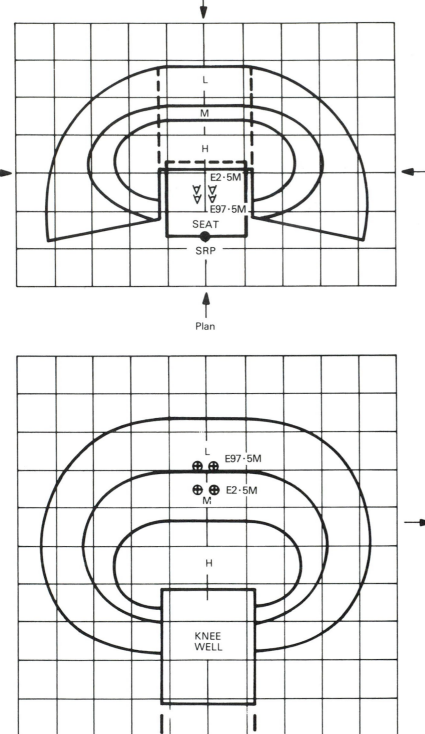

Plan

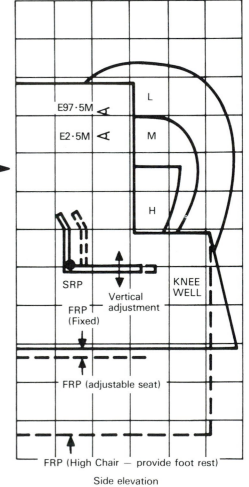

(d) SITTING REACH
suitable for a general work or high chair:
general UK adult males
(small – large) or average –
large females.

0·2 m grid.

*See Key, notes on use.

Rear elevation

Side elevation

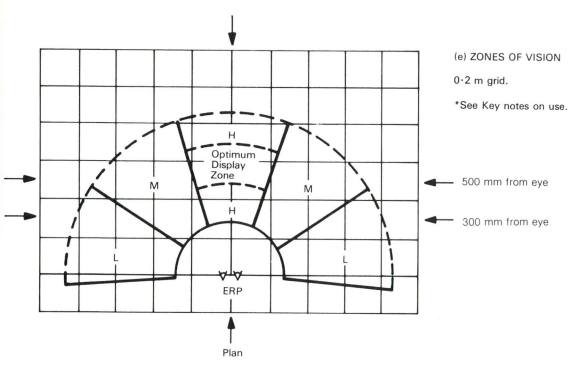

(e) ZONES OF VISION

0·2 m grid.

*See Key notes on use.

← 500 mm from eye

← 300 mm from eye

Plan

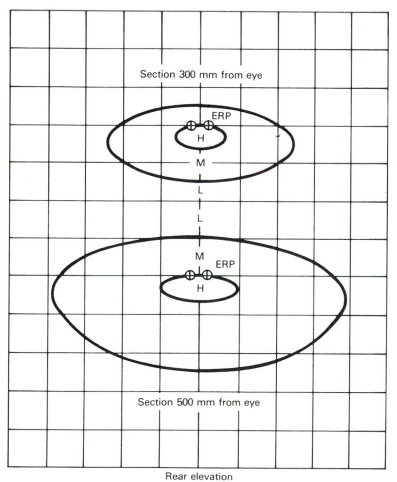

Section 300 mm from eye

Section 500 mm from eye

Rear elevation

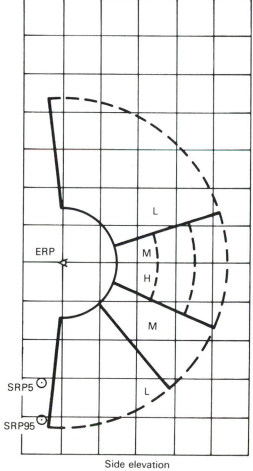

Side elevation

2.4 Clearance, access, size and safety distance details

2.4.1 Introduction

The following is intended as useful information to be used with the overlays.

These tabulated data provide recommended dimensions They are not exhaustive of all possibilities but deal with general cases.

The dimensions will need to be tested to see how adequate they are in practice (see Section 2.5, *Fitting trials*).

Reach dimensions all assume a grip with a full fist, not fingertip dimensions.

If the worker is to sit or stand, use knee and foot clearances from the sitting section with heights increased to set the work-surface at standing height. Provide a footrest at floor height as specified in the 'seated worker' section.

If the work height is greater than 50 mm from the recommended and the equipment cannot be changed, provide a false floor.

2.4.2 Clearance

Figure 2.4 Clearance and access dimensions for a standing worker.

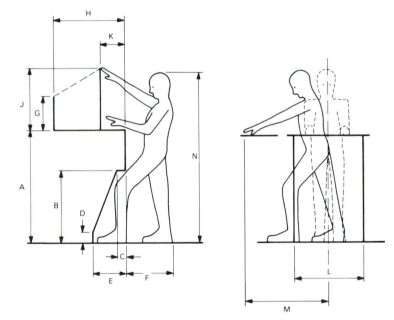

		%ile men (in mm)			%ile women (in mm)			
Code	Dimension	5	50	95	5	50	95	Comments
A	Work surface height (elbow)	970	1070	1190	905	995	1065	
B	Knee clearance height			560				Minimum required clearance
C	Knee clearance depth			125				
D	Toe clearance height			100				
E	Toe clearance depth			250				
F	Back clearance			915				
G	Max. reach access height	380			380			
H	Max. reach access distance	915			890			
J	Upright reach access height	685			635			
K	Upright reach access distance	455			405			
L	Crouch space for side pick-up			1145				
M	Extent of max. side reach[a]	1145						
N	Standing workers height	1600	1730	1880	1505	1605	1695	

[a]For a twisted trunk, M will be less than stated.

Figure 2.5 Clearance and access dimensions for a seated work.

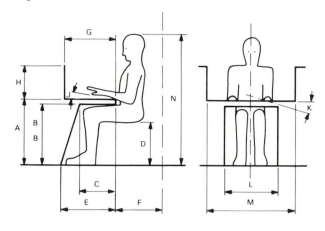

Code	Dimensions	%ile men (in mm)			%ile women (in mm)			Comments
		5	50	95	5	50	95	
A	Work surface height	660	710	760	635	685	735	
B	Knee clearance height (cross-legged)			750			725	Minimum required clearance
B'	Knee clearance height (un-crossed)			625			600	Minimum required clearance
C	Knee clearance depth			350			350	
D	Seat height (horizontal seat)	447	483	523	424	460	495	Use 5th %ile if only using min. B
E	Foot clearance depth			650			650	
F	Clearance for getting up			635			635	
G	Grasp objects depth	610			610			Assumes slight leaning
H	Grasp objects height	380			380			Assumes slight leaning
J	Keyboard or switch panel angle		15°			15°		
K	Keyboard angle for one-handed operation		20°			20°		
L	Knee-well width			650			650	
M	Distance apart of side panels for max. reach to grasp to height H	1650			1500			
N	Max. head height			1450				

2.4.3 Access

Table 2.2 Access clearance and size details.

		Height (mm)	Width (mm)	
WORKSTATION	See *Overlays* for seated and standing workspace clearances			
ADDITIONS TO DIMENSIONS	The following are minimum recommended dimensions. Extra allowances should be made for equipment, tools, removing components etc.			
GANGWAYS, AISLES	Minimum (moving sideways) Feeder aisles 2 people passing (one sideways) 2 people passing Catwalks	1000 (stoop) 2000 (walk)	330 510-765 765-915 1200-1375 Base 370 Top 635	
	Crawling	800	515	
	Prone	440	515	
LADDERS, STAIRS	Use ladder if rise > 50° or stairs impracticable Rise 75-90°. Provide guard rail at top. Ladder with rungs 32-38 mm dia. Rise 50-75°. Ladder with treads of depth 75-150 mm Rise 50°. Use stairs (opt. rise 30-35°) Rise <15°. Ramp may be used	 500 min Rise 125-200 Going depth 240-265		
	Provide handrail 850-900 mm above stairs		As for aisles	
OPENINGS, HOLES	Empty hand, flat Hand plus screwdriver Fist, clenched Inserting box with hands Thumb plus 2 fingers Hole for reaching to 150 mm (hands clasped) Hole for reaching to 645 mm Full arm reach	60 100 110 Box size + 45 50 105 105 ›100	105 100 110 50 115 460 600	
HANDLES	Provide handles, especially for components weighing over 4·5 kg	CLEARANCE, C MIN 50 OPT 65 (GLOVED) RAD, R 3 mm (< 7 Kg) — 10 mm (> 9 Kg) SEPARATION, S MIN 50 mm WIDTH, W ONE HAND 110 (bare) — 120 (gloved) 2 HANDS 220-240		

2.4.4 Manual handling of loads

Load handling

Current guidelines on handling loads differ between the European Union and the US. The EU requires hazard assessment on ergonomics lines (EC Manual Handling Directive 90/269/EEC).
The H&SE Manual Handling of Loads (1993) gives recommendations and guidance for assessment. This must take account of the characteristics of the load (e.g. weight; size; balance), the physical effort required (e.g. requires twisted trunk), the characteristics of the working environment, (e.g. distance carried; rate of working; frequency).

The US uses a revised NIOSH formula (Waters, 1993) which requires calculations taking account of six factors. These are horizontal and vertical start and finish points of the lift; vertical travel distance of load; assymetry of load; frequency of lift, and quality of handholds. The maximum load which may be lifted infrequently, with no vertical displacement, good grip with both hands, close to body and very short travel distance, is 23 kg for young males according to this formula.

Manual handling guidelines for design

1. Design on the assumption that appropriate manual handling equipment will be used; specify this on product.
2. If manually handled, design so that handler can do the task standing without bending or twisting.
3. Position objects to be handled so that the human body has space around when accessing the object or moving it.
4. Provide for objects to slide rather than having to be lifted; provide stops to give safe sliding limits.
5. Provide good handholds for grip in palm of hand; allow object to be held close to the body, elbows against sides of body.
6. Label objects with weight; mark handholds and slinging points. Indicate if centre of gravity is asymetric to casing.
7. Provide openings wide enough for seeing object when reaching for it, (see Section 2.4.3).

Factors affecting ability to handle loads

Weight of load.
Dimensions.
Horizontal distance of the load from the lifter.
Distance the load is moved from its origin to its destination.
Height the load is lifted vertically
from its origin to its destination;
height of the origin from the floor.
1- or 2-person lift.
Mechanical aids.
Frequency of handling.

Weight

1 person: Maximum for males = 20 kg.
           Maximum for females = 14 kg.
2 people: Not more than 32 kg, 22 kg for women. Label with weight and/or that load needs 2 people. If lifted over 1500 mm high and weighing over 16 kg (for men), use a mechanical lift.

Mechanical lifts must be used for loads and heights greater than specified, and warning labels and lifting eyes or other points must be provided.
Note that usable strength is related to many factors, including age, posture, fitness, training and experience, duration of effort and frequency of lift.
In general, reduce all loads by 1/3 for people over 40.
If weight, vertical distance and frequency of the lift are known, maximum *acceptable* weights of lift, for two widths of load and for two heights of lift, are shown in Figure 2.6. The graphs assume a load with good handles, able to be reached and held close to the body and not carried horizontally. They are also for people up to 40 years; reduce the loads by 1/3 for those between 40 and 60.

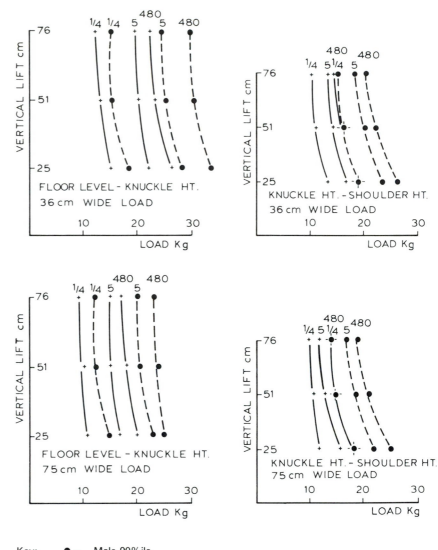

Figure 2.6 *Maximum acceptable weight of lift for 90% of American adult population showing effect of height of lift, frequency of lift, width of object, male and female differences (Snook, 1978).*

Key: ● = Male 90%ile
      + = Female 90%ile
¼, 5, 480 = Interval between lift in mins

When loads are manually handled, avoid designs which would require the lifter's trunk to rotate to reach or place the load. Very high and dangerous back loads can be created where lifts are not symmetrical in front of the body.

Repetitive work and upper limb disorders

Avoid designing work activities and equipment which require frequent repetitions of the same hand positions, grips or forces, or the repeated or constant holding of the upper arms away from the body. Relevant guidelines and legislation for the UK and EU are Health and Safety Executive (1990) *Work-related Upper Limb Disorders*. See also Putz-Anderson (1988).
Design should endeavour to minimize:
Deviations of the hand from being in a straight line with the wrist.
Repeated gripping, particularly with a wide spread of the fingers, by pinching or by pressing the fingers and thumb together.
Upper arm raising, frequent use of hands substantially above or below elbow level.
Sudden exertions, rapid actions or shock loading.

Continued maintenance of a posture for more than a few minutes without rest.
Continuously repeated short cycle activities throughout the day.
Frequent forceful squeezing, as in the cutting of thick wires or the use of scissors.

2.4.5 Safety distances

General

These correspond to a reach or body dimension plus a safety allowance, and are measured from points which are accessible for operating, maintaining and inspection.
The aim is to maintain hazard points (squeezing, shearing, cutting, etc.) at a safe distance from the operator.
The following dimensions are extracted from DIN 31001 part 1 (dimensions are for adults and in mm). See also BS EN 294 (1992).
Some information on openings in guards is provided to BS 5304 (1988) (*Code of Practice for Safety of Machinery*) figure 6, BS EN 292 (1992) Parts 1 and 2 and also BS 3042 (1992) (standard test fingers and probes).

Figure 2.7 Safety distances for whole-body reach.

Reaching up
With the body upright and standing at full height, the safety distance when reaching up is 2500 mm.

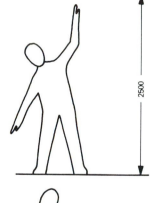

Reaching down, reaching over
When reaching down over an edge, e.g., on machine frames or safety features, the safety distance is found from:
A — Distance of hazard point from floor
B — Height of edge of safety feature
C — Horizontal distance of edge of hazard point

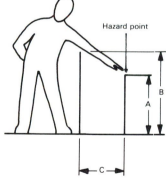

Distance of hazard point from floor A	Height of edge of safety feature B[a](in mm)							
	2400	2200	2000	1800	1600	1400	1200	1000
	Horizontal distance C from hazard point							
2400	—	100	100	100	100	100	100	100
2200	—	250	350	400	500	500	600	600
2000	—	—	350	500	600	700	900	1100
1800	—	—	—	600	900	900	1000	1100
1600	—	—	—	500	900	900	1000	1300
1400	—	—	—	100	800	900	1000	1300
1200	—	—	—	—	500	900	1000	1400
1000	—	—	—	—	300	900	1000	1400
800	—	—	—	—	—	600	900	1300
600	—	—	—	—	—	—	500	1200
400	—	—	—	—	—	—	300	1200
200	—	—	—	—	—	—	200	1100

[a] Values for edge B under 100 mm have not been included because the reach does not increase any further and in addition there is the risk of falling into the hazard area.

Reaching round

This covers the safety distance of freely articulating body parts around edges in any position, for adults and children.

The radius of movement about a fixed edge is determined by the reach of given body parts. The safety distances assigned below must be respected as minima if the body part concerned is not to be allowed to reach a hazard point. Of special importance is the hazard area which can be reached when these body parts are introduced through slots.

When applying safety distances it is to be assumed that the basic joint component of the relevant body part is in fixed contact with the edge. The safety distances apply only if it is ensured that further advance or penetration of the body part towards the hazard point is excluded.

Figure 2.8 Safety distances for reach of body parts.

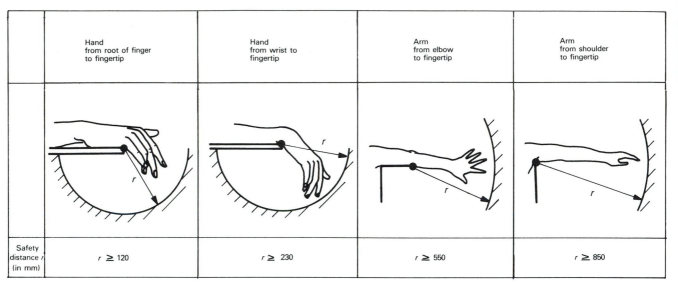

	Hand from root of finger to fingertip	Hand from wrist to fingertip	Arm from elbow to fingertip	Arm from shoulder to fingertip
Safety distance r (in mm)	$r \geq 120$	$r \geq 230$	$r \geq 550$	$r \geq 850$

Figure 2.9 Safety dimensions for elongated apertures with parallel sides.

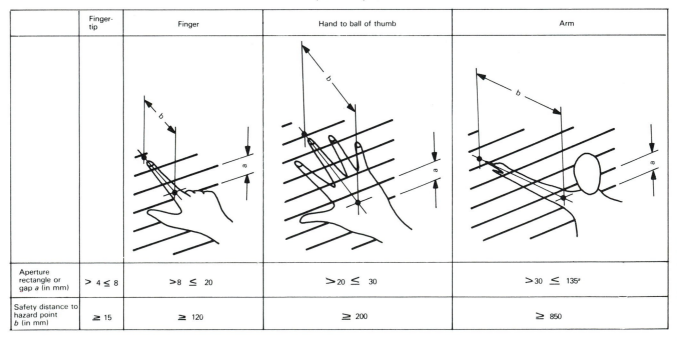

	Finger-tip	Finger	Hand to ball of thumb	Arm
Aperture rectangle or gap a (in mm)	$> 4 \leq 8$	$> 8 \leq 20$	$> 20 \leq 30$	$> 30 \leq 135^a$
Safety distance to hazard point b (in mm)	≥ 15	≥ 120	≥ 200	≥ 850

[a] With measurements exceeding the stated aperture it is possible for the body to stoop in, so that the safety distances in accordance with Section 2.4.3, p. 40 are to be taken into account.

Figure 2.10 Safety dimensions for square or circular apertures.

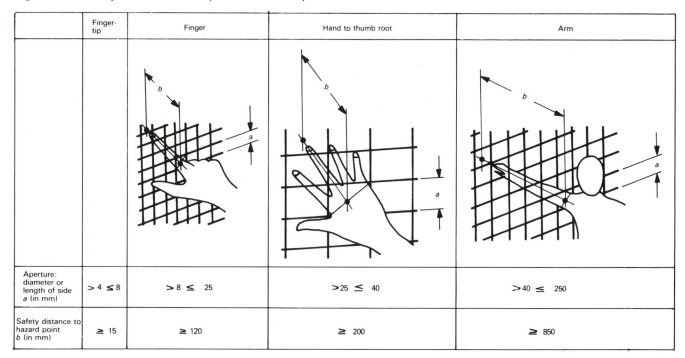

	Finger-tip	Finger	Hand to thumb root	Arm
Aperture: diameter or length of side *a* (in mm)	> 4 ≤ 8	> 8 ≤ 25	> 25 ≤ 40	> 40 ≤ 250
Safety distance to hazard point *b* (in mm)	≥ 15	≥ 120	≥ 200	≥ 850

Figure 2.11 Safety dimensions at squeeze points.

A squeeze point is not regarded as a hazard point for the indicated body parts if the safety distances are not less than shown below and if it is ensured that the next larger body part cannot be introduced:

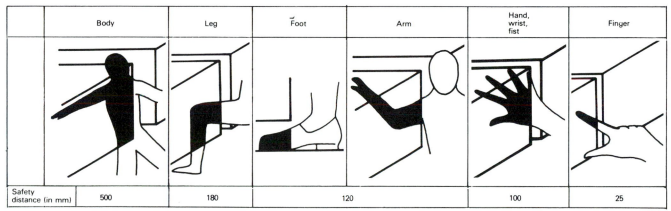

	Body	Leg	Foot	Arm	Hand, wrist, fist	Finger
Safety distance (in mm)	500	180	120		100	25

2.5 Fitting trials

2.5.1 General

The use of the following approach is recommended for confirmation of preliminary designs and first approximations (see *Limitations of data*, pp. 28–29).

Full trials are recommended or, if time is limited, abbreviated procedures are valuable.

List all user actions and carry out trials for all main actions affecting dimensions and layout.

Users should be representative in terms of size ranges, training, etc.

Evaluations are more satisfactory when two or more alternatives are compared. Where one proposed design is tried, subjects' performances and judgements must be compared with an assumed normal unless a fairly simple design decision is being tested. (The design is acceptable if the subjects do better or no worse than the assumed normal.)

2.5.2 Simulator

May be simple (e.g., cardboard) for quick checks of simple problems, or more complex (slotted angles, softboard, etc.) for more important problems or detailed evaluations.

Each significant dimension should be adjustable without altering other significant dimensions.

Quick alteration is essential (so that memory can be used in comparison).

Set up the simulator according to a first approximation based on design principles, check charts, etc.

2.5.3 Designer as subject

This is an opportunity for the designer to systematically decide on standards of convenience and comfort for each operator action.

The designer reproduces the operator's actions and adjusts each significant dimension in turn to establish a range of settings compatible with the appropriate range of body sizes and assessments of convenience and comfort.

The importance of each dimension and maximum/minimum limiting points are established.

2.5.4 Abbreviated method

If full trials are not possible estimate limiting settings for extreme sizes:

$$\text{Extreme subject's setting} = \text{Designer's setting} \times \frac{\text{Extreme subject's height}}{\text{Designer's height}}$$

Omit all but the most critical maxima/minima.

2.5.5 Full trials

Subjects

Try out with the designer as the subject, and then with a range of subjects.

Ensure that the subjects are representative in body size (see *Body-size variations*, p. 26).

Arbitrarily cut off the extremes (e.g., 5th and 95th %iles; see *Body-size variations*, p. 26).

Pick fairly tall and fairly short subjects. Use tall fat and short fat subjects if possible, as they are less adaptable than thin people – if there is no discomfort for fat people then there will not be for thin either.

Subjects should be of an appropriate age, background and skill.

Allow for attitude, practice, tiredness and individual differences.

Testing

Determine the range of each of the dimensions to be tested which is acceptable to each subject.

Set all dimensions to an average position.

The subject carries out one of the operator actions. The dimension to be tested is adjusted with the others constant.

Adjustment begins with the dimension at an initial setting much below the acceptable minimum. Adjustment is increased in steps (e.g., 25 mm) until it is above the acceptable maximum, and then starting well above this maximum, is reduced again to below the acceptable minimum.

The subject compares the comfort of each setting with the previous setting, or each setting with an estimate of what is an intolerable degree of discomfort or inconvenience (e.g., 'is this better' or 'is this tolerable').

It is thus possible to obtain minimum and maximum tolerable and optimum settings for increasing and decreasing adjustments.

Tolerance ranges are obtained for each dimension for each subject and plotted.

The ranges are found at settings not outside the tolerance ranges of any subject.

Figure 2.12 The tolerance range of dimensions.

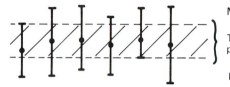

Maximum setting

Tolerance range for all subjects and most preferred best setting for each subject

Minimum setting

If there is no overlap, redesign or make adjustable.

If there are incompatible combinations adjust all dimensions within the final range or find combinations of settings involving the smallest departure from the intended degree of comfort and convenience.

3 Environmental Design

ENVIRONMENTAL DESIGN

3.1 Introduction

3.1.1 General arrangements

The ergonomic criteria for a good environment are those which will aid people in achieving their objectives whilst retaining effort, stress and errors within tolerable limits. The environment should be designed to help people, rather than just to remove its more objectionable or unacceptable features.

This section covers four of the six environmental factors:

Thermal;
Visual;
Auditory;
Vibrational.

Figure 3.1 Major environmental factors in workspace design.

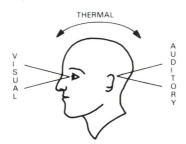

The two factors not covered are:

Chemical;
Radiation.

Each environmental factor will be considered in isolation. The components of each factor combine to cause the sensation and responses to the factor, whether or not it is safe, acceptable, etc. At the end of each section texts are recommended to provide more information, particularly concerning the complex interactions between factors.

3.2 The auditory environment

3.2.1 Basic principles

Noise is unwanted sound.
Legally the average level of noise should not exceed 85 dB(A) during the length of a working day to protect 90% of the exposed population from suffering deafness. But noise also hinders performance, hence attention must be paid to other facets of the noise and to the requirements of the job.

3.2.2 The effects of a poor auditory environment

Negligent design of auditory environments can:
Inhibit speech communication;
Mask warning signals;
Reduce mental performance;
Induce nausea and headaches;
Induce tinnitus (ringing in the ears);
Temporarily impair hearing;
Cause temporary deafness;
Cause permanent deafness.

3.2.3 The main factors in the auditory environment

Intensity

Intensity is the sound level or loudness: the pressure of the sound waves.
The loudness is measured as the ratio of the sound pressure to that of the pressure for a just-audible sound. The ratio is logarithmic, to enable the enormous range of audibility to be expressed in convenient numbers.

Speech Interference Levels (SILs) will give guidance on the noise levels which interfere with speech communication.

Unit of intensity $= N\,m^{-2}$.
Unit of loudness (i.e., the human perception of intensity) $= dB(A)$.
Noise intensity for an 8 h exposure should not exceed 85 dB(A).
Very short exposures should not exceed 135 dB(A), except for impulse noise whose instantaneous level should never exceed 150 dB(A).

Two noises must have a difference in intensity of about 10 dB(A) before they can be separated by the ear.
A change of 3 dB(A) means doubling the physical effect of the noise; small changes in level are thus important.

Figure 3.2 Noise levels which barely permit reliable conversation.

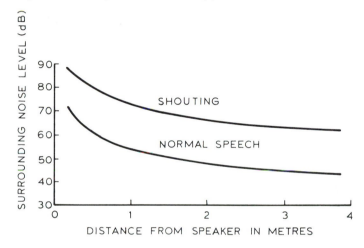

Frequency

Frequency, or pitch, is the rate of repetition of the cycles.
Human perception of frequency (in the audible range) varies with loudness.
In the normal range of industrial noise human hearing is more sensitive to the higher frequencies. If high frequency noise is present, then shorter exposure times or lower intensities are required for equivalent exposure effects.

Unit of frequency = Hertz (Hz)
(1 Hz = 1 cycle per second).
High frequencies mask lower frequencies.

Fluctuating frequency noise can be heard over a steady pitch.

Exposure time

Exposure time is the maximum time unprotected ears may be exposed to different intensities and frequencies of noise.

This example is from Swedish Standards (SEN 590111) at 1000 Hz.

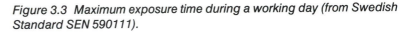

Total exposure during one day	Maximum intensity (dB(A))
>5 h	85
5–2 h	90
2–1 h	95
<20 min	105
<5 min	120

Figure 3.3 *Maximum exposure time during a working day (from Swedish Standard SEN 590111).*

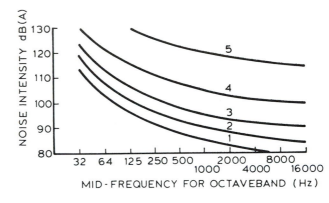

1. >5 hours
2. 2–5 hours
3. 1–2 hours
4. <20 mins
5. <5 mins

Where the intensity cannot be controlled at source, ear defenders and ear plugs can lengthen the permissible exposure period depending on the sound attenuating characteristics of the equipment, but a regular audiometric survey programme is desirable.
Ear defenders and ear plugs reduce *all* sound, including warning signals and speech.

3.2.4 Infrasound

Infrasound is very low frequency sound. Research has shown that at the lower end of the sound spectrum, 16–3·5 Hz, subjects have complained of the sensations of vertigo, nausea and headaches.

3.2.5 Ultrasound

Ultrasound is very high frequency sound. There is insufficient evidence to support the case that ultrasound induces sickness.

3.2.6 Legal requirements

The legal requirements are laid out in *Noise and the Worker*, by Her Majesty's Factory Inspectorate (1971).

BS 7445 Parts 1,2 and 3 (1991). Also ISO 1996, Part 1 (1982) and Parts 2 and 3 (1987).

3.2.7 Non-legal requirements

Just meeting the legal requirements is often misleading; they are only the maximum limits for minimizing deafness.
Noise within the legal limits can still have a very disturbing effect, especially on mental performance. For a general appreciation of the influence of sound see Burns (1973).

3.3 The thermal environment

3.3.1 Basic principle

To provide an environment which imposes as small a load as possible on the body's own thermoregulatory system.
The heat exchanges between the body and its environment can be simply represented by the heat balance equation:

$$M \pm C_1 \pm C_2 \pm R - E = S$$

where M is the heat due to metabolism, including physical work effort; C_1, C_2 and R are the heat lost or gained by convection, conduction and radiation; E is the heat lost due to the evaporation of sweat; and S is the amount of heat lost or gained by the tissues of the body. If the body maintains thermal equilibrium, then $S = 0$.

3.3.2 The effects of a poor thermal environment

Negligent design of thermal environments can cause:
Reduced physical performance;
Irritability and distraction from the task, and reduced mental performance;
Discomfort from sweating or shivering;
Increased load on the heart;
Death.

ENVIRONMENTAL DESIGN

53

3.3.3 The main factors in the thermal environment

Activity level

The activity level affects metabolism. Physical activity generates body heat, which the environment must compensate for.
High workloads require cooler environments.
Low workloads require warmer environments.

Activity level unit = kcal h^{-1} m^{-2} (heat equivalent per hour per square metre of body surface).

An example from Fanger (1970).

Activity	Metabolism (kcal h^{-1} m^{-2})
Draughtsman	60
Light machine work	100–120
Heavy machine work	200
Slag removal in a foundry	380

Clothing

If special clothing is worn as part of the job this will change the heat loss equation.
Remember that people's dress habits change seasonally *regardless* of the indoor environment.

Clothing unit = clo values or a verbal description, i.e., light, medium, winter, etc.

An example from Fanger (1970)

Clothing	Clo factor
Nude	0
Light summer clothes (male)	0·5
Heavy business suit	1·5
Polar weather suit	3–4

Ambient temperature

Ambient temperature (also known as air temperature or dry bulb temperature) is simply the temperature of the surrounding air. Air temperature alone is *not* a sufficient measure of the suitability of an environment.
Here are some examples of the order of temperatures for different types of job.
Ambient temperature is measured by a conventional mercury in glass thermometer (dry bulb).

An example from Bell (1974).

Activity	Temperature (°C)
Clerical work	20·0–19·5
General office work	19·5–18·3
Active workers in light industry	18·3–15·5
Heavy industry	15·5–12·8

Humidity

Humidity (or relative humidity) is the water content of air.
Humidity can vary over a wide range without much effect in normal working conditions.
Humidity is critical in a hot environment where it will restrict heat loss by evaporation.
Humidity is measured by using wet and dry bulb thermometers and hygrometric tables.

Humidity units = % water saturation of air.

An example from Bell (1974). At 18·5°C for general office work humidity could range from 30 to 70% with little change in human comfort.

ENVIRONMENTAL DESIGN

Air flow	Air flow (or air speed) is the velocity of air at the individual workplace. Air flow is important for cooling and for the sensation of fresh air. Air flow is measured by an anemometer.	Air flow unit $= \mathrm{m\ s^{-1}}$. An example from Bell (1974). For an office a mean air flow of $0.11–0.15\ \mathrm{m\ s^{-1}}$ would be judged comfortable. $0.5\ \mathrm{m\ s^{-1}}$ is judged uncomfortably draughty. The heavier the work, the higher the acceptable rate of air flow.
Radiant temperature	Radiant heat is the heat energy transferred *to or from* the body through radiation. In most cases it is the radiant temperature transmitted to the body which is of interest. Where the radiant temperature exceeds the ambient temperature by 10°C or more the sources should be shielded. Where the operator is given protective clothing to reduce the effect of radiant heat a new microclimate is created and the normal processes of heat loss are severely restricted. Hence this may increase discomfort and reduce work capacity due to the heat gain within the clothing. Radiant temperature is measured by a mercury in glass thermometer within a blackened copper ball, a 'globe thermometer'.	Radiant temperature unit = °C. An acceptable radiant temperature range is 16°–20°C. An unduly cold or warm surface, by accepting or emitting radiant heat, affects feelings of comfort. Note that increased air flow does not compensate for radiant heat exposure: shielding is necessary.

3.3.4 Techniques to evaluate the thermal environment

Thermal stress	The international standard scale (ISO 7730, 1984) to evaluate thermal stress is the wet bulb globe temperature, WBGT.
Thermal comfort	A widely accepted scale for thermal comfort in more moderate environments is Fanger's Thermal Comfort Index.
Cold environments	No standard reference exists.

3.3.7 Legal requirements

The legal requirements are set out in the Factories Act 1961.
The legal requirements are the minimum requirements and do not imply that the conditions are optimal.
Also relevant is ISO 7730 (1984) *Moderate Thermal Environments – Determination of the PMV and PPD Indices and Specification of the Conditions for Thermal Comfort* (see Fanger 1970).

PMV = Predicted mean vote
PPD = Predicted percentage dissatisfied.

3.4 The visual environment

3.4.1 Basic principle

The aim of designing visual environments is not to provide light but to allow people to recognize what they see. (NB The visual environment is the most complex of the three to design.)

3.4.2 The effects of a poor visual environment

Negligent design of visual environments can induce:
Visual discomfort and headaches;
Errors and inability to see detail;
Confusion, illusions and disorientation;
Epilepsy (where it is already present).

3.4.3 The main factors in the visual environment

Illuminance

Illuminance is the amount of light *falling* onto a surface.
The amount of illumination required to light a task adequately depends on the task.
Illuminance falls off as the square of the distance from the source.
Recommended levels can be found for almost all tasks in the CIBS code.
Illuminance is measured by a light meter at the work surface.

Illuminance unit = lux (SI).

An example from the CIBS Code (1984).

Task	Illuminance (Lux)
Loading bays, storerooms Packing work, mould preparation, general engineering	150–300
engineering	300–500
Office work, fine engineering, inspection and steel works	500–800
Drawing office, garage and tool room, watchmaking	> 800

Luminance

Luminance is defined as the amount of light emitted by a surface.

A luminance ratio of 10:3:1 between the task to the surrounding area to the general background has been found to be comfortable.

Excessive luminance causes glare. Insufficient luminance reduces visibility.

Luminance unit = candela m^{-2} (SI).

See *Contrast* for formula.

Concentration is helped if the work area is the brightest part of the visual field.

Contrast

An object can be seen and its shape identified because of its contrast with its background.

Contrast can be improved by:
Changing the level of illumination;
Changing the reflectivity of certain parts of the task;
Using directional lighting to cast shadows.

There is no agreed measure of contrast. The most common method is

$$C = (L_1 - L_2)/L_1$$

where C is the contrast or luminance ratio; L_1 is the brighter of
the two luminances; and L_2 is the lower of the two luminances.

Glare

Glare occurs when there are areas of high brightness in the visual field.

There are two main types of glare (both must be avoided):
Discomfort glare, i.e., glare which causes discomfort only;
Disability glare, which causes discomfort and a drop in visual performance by reducing the ability to see detail.

Glare from a single source is measured and expressed as a glare constant (see Hopkinson and Collins, 1977).

This is then converted to a Glare Index.

Where there is more than one glare source at a workplace, they can be summed to give the Glare Index.

An example from CIBS code (1984).
(*a*) Environments where no glare is permissible:
Glare Index Limit = 10.
(*b*) Environments where glare must be kept to a minimum:
Glare Index Limit = 13.
(*c*) Environments where different degrees of glare are permitted depending on the sensitivity of people, the time spent in a room and the attention demanded by work:
Glare Index Limit = 16–28.

ENVIRONMENTAL DESIGN

Flicker

Flicker arises from poor quality fluorescent fittings or rotating parts between a light source and the eye.
If it cannot be eliminated, arrange that the flicker is greater than the Critical Flicker Fusion Frequency (CFF) of people for the level of field luminance. The CFF is the frequency at which flicker becomes imperceptible. This threshold varies greatly between people but can be as high as 85 Hz, i.e., higher than mains frequency.
Also, the greater the luminance level the greater the human sensitivity to flicker.

Flicker can be noticed more easily at the edge of the field of vision. Distribution of lights across the three phases of the power supply, and the choice of phosphors in fluorescent lights, can eliminate flicker.

Colour

Artificial lighting has colour. Choosing the colour of lighting is important emotionally and with respect to colour.

Coloured objects look white under light of their own colour and black under light of a complementary colour, e.g., red objects look white in red light and black in green light.

1. Emotionally.

Verbal impressions associated with different levels and colours of fluorescent lighting in a conference room.
An example from Bodman (1967; cited in Hopkinson and Collins, 1977).

Average illumination (lux)	Colour of light		
	Warm white	White	Daylight
700	Not un pleasant	Dim	Cool
700–3000	Pleasant	Pleasant	Neutral
3000	Exces sive, Artificial	Pleasant, lively	Pleasant

2. With respect to performance. Where colour is important in the task to indicate danger or the quality of goods (i.e., inspection), select lighting which increases the contrast of the colour to be recognized against the background.

The visual state of the operator

If complaints occur about the state of the visual environment or about visual performance, consider the possibility of examining the visual state of the operator.

3.4.4 Legal requirements

The legal requirements are set out in *Lighting in Offices, Shops and Railway Premises* (Her Majesty's Factory Inspectorate, 1978). Note that the legal requirements are lower than the CIBS (1984) recommendations. Where possible follow the CIBS Code of Practice. The way the factors which make up the visual environment interact is complicated and requires the advice of a lighting engineer.

3.5 The vibratory environment

3.5.1 Basic principles

Vibration can produce noise (see Section 3.2) and when transmitted to the body, can produce discomfort, difficulties in operating equipment, sickness and injury. Investigations commonly deal with whole body vibration or hand–arm vibration. Effects accumulate over time for both types of exposure. Body resonances (amplification of certain imposed frequencies) can increase the likelihood of body damage.

3.5.2 The effects of a poor vibration environment

Whole body vibration: at low levels interferes with control movements; reduces readability of displays; causes motion sickness; accelerates fatigue from muscular efforts to control body movements and from difficulties in breathing; increases risk of back-pain.
At high levels causes injury to internal organs; increases risk of falls and collision injuries.
Hand–arm vibration: introduces finger-tingling and numbness; causes vibration white finger (VWF) the effects of which are exacerbated by cold.

3.5.3 The main factors in the vibration environment

Frequency (see Section 3.2.3). Physiological effects occur at various frequencies. Increase mass to lower frequency; introduce isolators to separate handles or vibrating component from rest of machine; stiffen components which are vibrating; improve balancing; increase damping; require high standards of maintenance.
Body is most sensitive to vibration in vertical direction. Reduce vibration in frequency range of 1–4 Hz to reduce interference with breathing, 4–10 Hz to reduce internal injury, chest and abdominal pains and muscle reactions, up to 12 Hz to reduce back strains; head is affected at about 25 Hz and eyes up to 80 Hz.
Acceleration: acceleration is related directly to energy input; VWF more likely with higher accelerations. Relationship of frequency and exposure time limits acceptable for avoiding VWF are given in Figure 3.4 (overleaf), for ½, 1, 2, 4 and 8 hours of exposure per shift, lines 5, 4, 3, 2 and 1 respectively.
Legal requirements: VWF is a reportable industrial disease in many countries. BS 6841 (1987) for whole-body vibration and BS 6842 (1987) for hand–arm vibration are relevant.

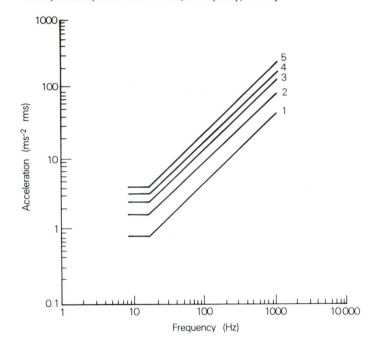

Figure 3.4 *Acceptable limits of vibration exposure of the hand (ISO 5349, 1986). Curves 1 to 5 refer to multiplying factors, associated with exposures per 8 hour shift of 4–8, 2–4, 1–2, ½–1, and up to ½ respectively. (From* Evaluation of Human Work, *Wilson, J.R. and Corlett, E.N. (Eds), 1995.)*

4 Control Design

CONTROL DESIGN

61

4.1 Design criteria

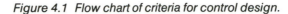

Figure 4.1 Flow chart of criteria for control design.

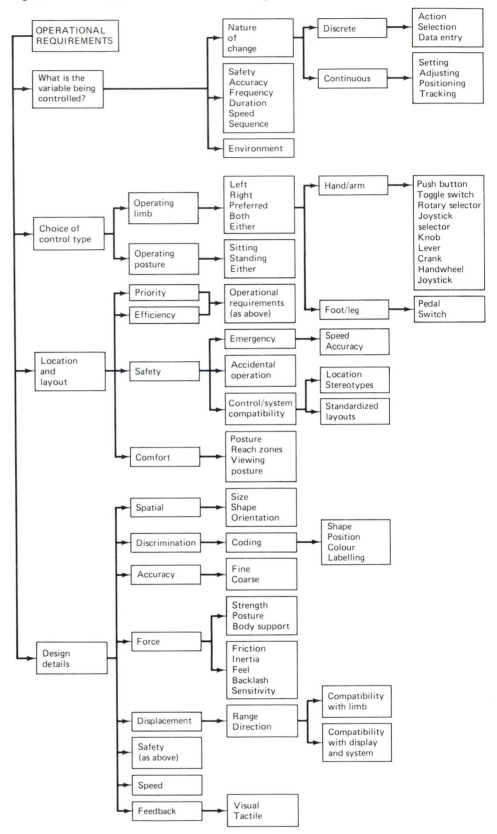

4.2 Choice of control type

Table 4.1 The general suitability of controls for different types of operation.

Control Type	High accuracy	High speed	Large force	Large displacement range	No. of discrete settings	High frequency of operation	Long duration of operation	Sequence	Visual identification	Non-visual identification	Check reading of position	Emergency action	Ease of compatibility with system response
Push button	—	H	—	—	2	H	L	H	L	L	L (Unless lit)	H	H
Toggle switch	—	H	—	—	2/3	M	L	M	M	H	H	H	L
Rocker switch	—	H	—	—	2/3	M	L	M	M	H	H	H	L
Rotary selector	H	H	—	—	3 – 24	M	L/M	M	M	M	M	M	H
Joystick selector	H	H	L	—	4 – 8	H	L	L	M	M	M	M	—
Cranks:													
Small	L	H	L	H								H	
Large	H	L	H	H	—	M	M	—	H	H	L	L	M
Horizontal	M	M	H	H									
Vertical	L	H	M	H									
Handwheels	L	H	L	H									
	H	L	L	L	—	M	M/H	—	H	M	L	L	M
	—	L	H	L									
Levers:													
Horizontal	L	H	L	L								M	L
Vertical (to/fro)	M	H	Short L	L									
	M	H	Long H	L	—	M	M	M	M	M	H	H	M
Vertical (across body)	M	M	M	—								M	L
Knob	M	—	—	M	—	M/L	M/L	H	M	M	L	—	M
	H		L	L									
Joystick	M	M	L	L	—	H	H	—	M/H	H	L	L	M
Pedals:													
Leg	M/L	M	H	—	—	M	M	L	—	M	—	H	M
Ankle	H	M/H	L	—	—	H	H	M	—	M	—	M	M
Footswitch	—	H	L	—	—	L	L	—	—	M	—	H	H

General suitability: H = High; M = Medium; L = Low; — = Unsuitable or not applicable
Note that high accuracy, high speed, large force and large displacement are generally incompatible.

4.3 Location and layout

4.3.1 General

Task requirements

Undertake task analysis to determine operational requirements. Arrange for efficient, safe and comfortable operation according to priority, frequency, duration, force, speed, accuracy and sequence and other basic requirements.

Reach

Locate high priority controls in the high priority zone in front of the operator (see *Workspace design*, p. 17). Ensure controls required during maintenance are accessible as appropriate (see Chapter 6, *Maintainability*, p. 101).

Limb

Divide between left and right hands and feet according to the need for simultaneous operation and variation in the preferred hand or foot (locate centrally for operation by right- or left-handed people).

Search

Controls to be in an expected or standardized layout with respect to each other.
More separation between hand controls is needed for 'blind' searching and location, to avoid accidental operation.

CONTROL DESIGN

Compatibility	Controls must be in the same spatial layout as the system or displays.
Standardization (of layout, etc.)	Important where the operator moves between different machines, or where training is limited.
Consistency	Displays must be in a consistent location, and must move in a consistent (compatible) direction relative to their controls.
Location coding	By the position in relation to other controls, as well as by shape, size and texture.

4.3.2 Grouping

Avoid interference.
Arrange in order of priority.

Function

Arrange controls of similar function together (dissociate if confusion is likely).

Rows and columns

A maximum of three vertical columns and three horizontal rows.

4.3.3 Sequence

Sequential operations must flow smoothly.

Place sequential controls in close proximity. If the sequence never varies, incorrect responses can be eliminated by interlocking.

Where controls are sometimes omitted from the sequence use location coding.

Operate left to right and top to bottom.

4.3.4 Control – display relationships

Controls must be close to the appropriate displays *or* grouped in similar patterns.

Controls and appropriate displays or work must be operable/visible simultaneously without adopting an uncomfortable posture.

The left-hand control of a row of controls must refer to the top display of a column of displays, etc.

The smallest of a set of stacked knobs must refer to the left-hand display. See also *Compatibility* (above).

CONTROL DESIGN

4.4 Design details

Figure 4.2 Design details of controls.

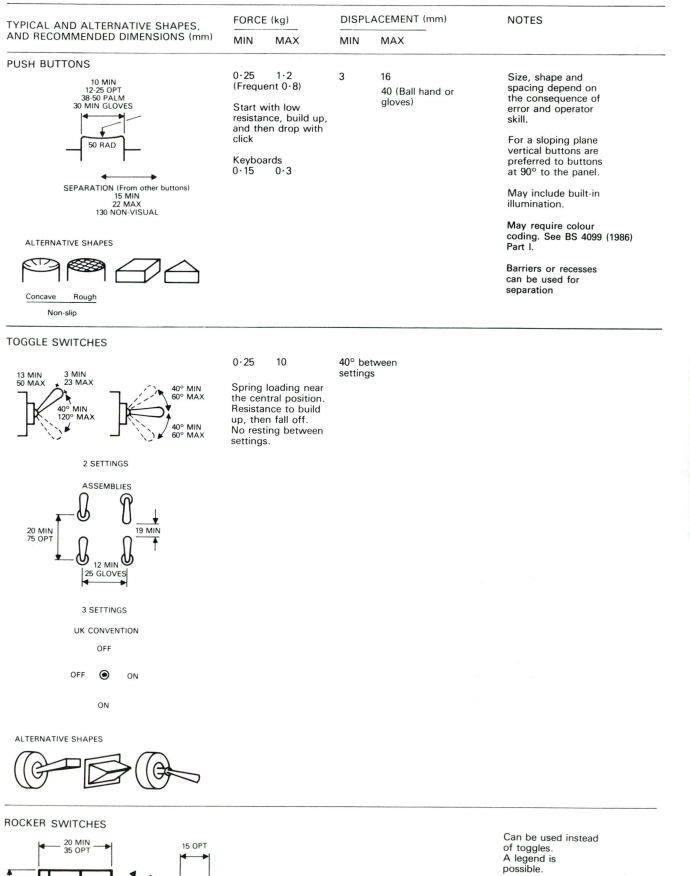

TYPICAL AND ALTERNATIVE SHAPES, AND RECOMMENDED DIMENSIONS (mm)	FORCE (kg)		DISPLACEMENT (mm)		NOTES
	MIN	MAX	MIN	MAX	
PUSH BUTTONS	0·25 (Frequent 0·8) Start with low resistance, build up, and then drop with click Keyboards 0·15	1·2 0·3	3	16 40 (Ball hand or gloves)	Size, shape and spacing depend on the consequence of error and operator skill. For a sloping plane vertical buttons are preferred to buttons at 90° to the panel. May include built-in illumination. May require colour coding. See BS 4099 (1986) Part I. Barriers or recesses can be used for separation

PUSH BUTTONS

10 MIN
12-25 OPT
38-50 PALM
30 MIN GLOVES

50 RAD

SEPARATION (From other buttons)
15 MIN
22 MAX
130 NON-VISUAL

ALTERNATIVE SHAPES

Concave   Rough
Non-slip

TOGGLE SWITCHES

TOGGLE SWITCHES	0·25	10	40° between settings		
	Spring loading near the central position. Resistance to build up, then fall off. No resting between settings.				

13 MIN
50 MAX
3 MIN
23 MAX
40° MIN
60° MAX
40° MIN
120° MAX
40° MIN
60° MAX

2 SETTINGS

ASSEMBLIES

20 MIN
75 OPT
19 MIN
12 MIN
25 GLOVES

3 SETTINGS

UK CONVENTION

OFF

OFF ⊙ ON

ON

ALTERNATIVE SHAPES

ROCKER SWITCHES

20 MIN
35 OPT
15 OPT
10 MIN
25 OPT
ON OFF
30°

Can be used instead of toggles. A legend is possible.

Figure 4.2 (cont.)

TYPICAL AND ALTERNATIVE SHAPES RECOMMENDED DIMENSIONS (mm)	FORCE (kg)		DISPLACEMENT (mm)		NOTES
	MIN	MAX	MIN	MAX	

ROTARY SELECTOR
(KNOB OR ROTARY BAR)

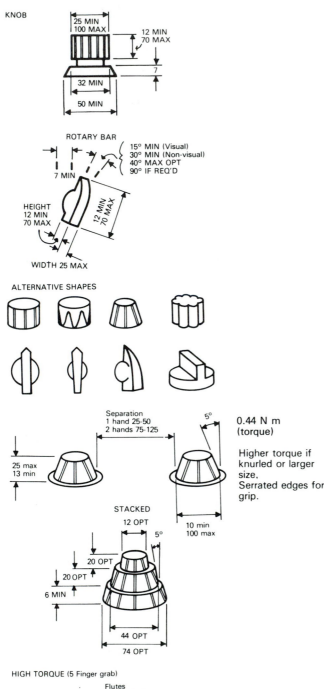

KNOB

25 MIN
100 MAX

12 MIN
70 MAX

7

32 MIN

50 MIN

ROTARY BAR

15° MIN (Visual)
30° MIN (Non-visual)
40° MAX OPT
90° IF REQ'D

7 MIN

HEIGHT
12 MIN
70 MAX

12 MIN
70 MAX

WIDTH 25 MAX

ALTERNATIVE SHAPES

A skirt gives a more visible scale and prevents damage to scale.

Bar knobs are preferable for checking settings in an array.

30° detents can be identified by feel.

Separation
1 hand 25-50
2 hands 75-125

5°

25 max
13 min

10 min
100 max

STACKED

12 OPT

5°

20 OPT

20 OPT

6 MIN

44 OPT

74 OPT

0.44 N m (torque)

Higher torque if knurled or larger size.
Serrated edges for grip.

Unlimited (depends on operating time)

Unsuitable for rapid adjustment unless folding crank handle

HIGH TORQUE (5 Finger grab)

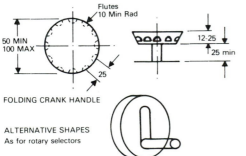

Flutes
10 Min Rad

50 MIN
100 MAX

12-25

25 min

25

FOLDING CRANK HANDLE

ALTERNATIVE SHAPES
As for rotary selectors

Figure 4.2 (cont.)

TYPICAL AND ALTERNATIVE SHAPES RECOMMENDED DIMENSIONS mm	FORCE (kg)		DISPLACEMENT (mm)		NOTES
	MIN	MAX	MIN	MAX	

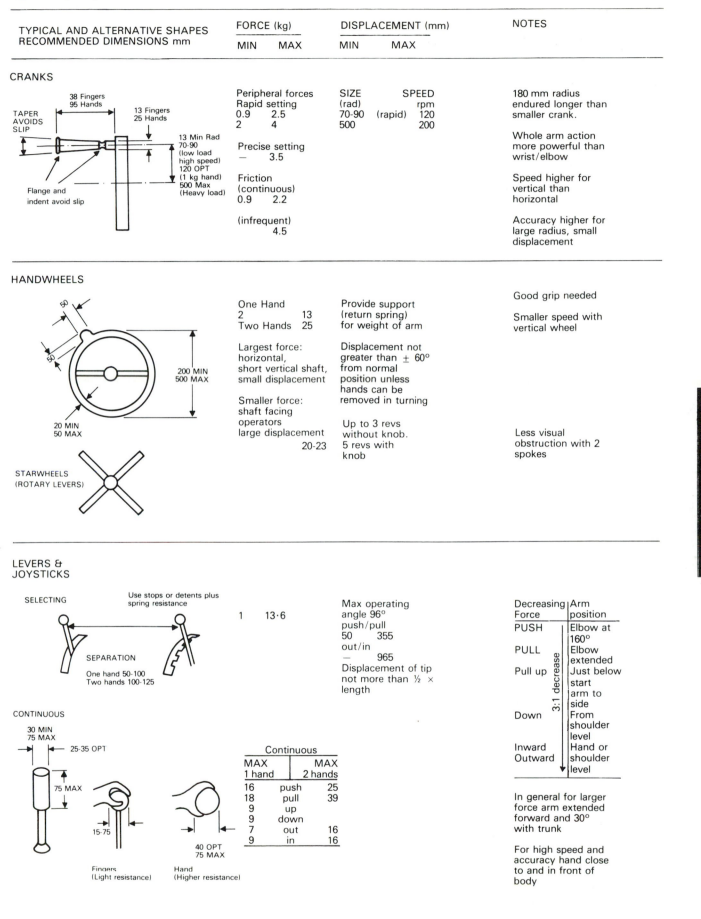

CRANKS

38 Fingers
95 Hands

13 Fingers
25 Hands

TAPER AVOIDS SLIP

13 Min Rad
70-90 (low load high speed)
120 OPT (1 kg hand)
500 Max (Heavy load)

Flange and indent avoid slip

FORCE:
Peripheral forces
Rapid setting
0.9 — 2.5
2 — 4

Precise setting
— 3.5

Friction (continuous)
0.9 — 2.2

(infrequent)
— 4.5

DISPLACEMENT:
SIZE (rad) — SPEED rpm
70-90 (rapid) — 120
500 — 200

NOTES:
180 mm radius endured longer than smaller crank.

Whole arm action more powerful than wrist/elbow

Speed higher for vertical than horizontal

Accuracy higher for large radius, small displacement

HANDWHEELS

50

200 MIN
500 MAX

20 MIN
50 MAX

FORCE:
One Hand
2 — 13
Two Hands — 25

Largest force: horizontal, short vertical shaft, small displacement

Smaller force: shaft facing operators large displacement
— 20-23

DISPLACEMENT:
Provide support (return spring) for weight of arm

Displacement not greater than ± 60° from normal position unless hands can be removed in turning

Up to 3 revs without knob.
5 revs with knob

NOTES:
Good grip needed

Smaller speed with vertical wheel

Less visual obstruction with 2 spokes

STARWHEELS (ROTARY LEVERS)

LEVERS & JOYSTICKS

SELECTING

Use stops or detents plus spring resistance

SEPARATION
One hand 50-100
Two hands 100-125

CONTINUOUS

30 MIN
75 MAX

25-35 OPT

75 MAX

15-75

40 OPT
75 MAX

Fingers (Light resistance)

Hand (Higher resistance)

FORCE:
1 — 13·6

Continuous		
MAX 1 hand		MAX 2 hands
16	push	25
18	pull	39
9	up	
9	down	
7	out	16
9	in	16

DISPLACEMENT:
Max operating angle 96°
push/pull
50 — 355
out/in
— 965
Displacement of tip not more than ½ × length

NOTES:

Decreasing Force	Arm position
PUSH	Elbow at 160°
PULL	Elbow extended
Pull up	Just below start arm to side
Down	From shoulder level
Inward Outward	Hand or shoulder level

(3:1 decrease)

In general for larger force arm extended forward and 30° with trunk

For high speed and accuracy hand close to and in front of body

Figure 4.2 (cont.)

TYPICAL AND ALTERNATIVE SHAPES RECOMMENDED DIMENSIONS (mm)	FORCE (kg)		DISPLACEMENT (mm)		NOTES
	MIN	MAX	MIN	MAX	

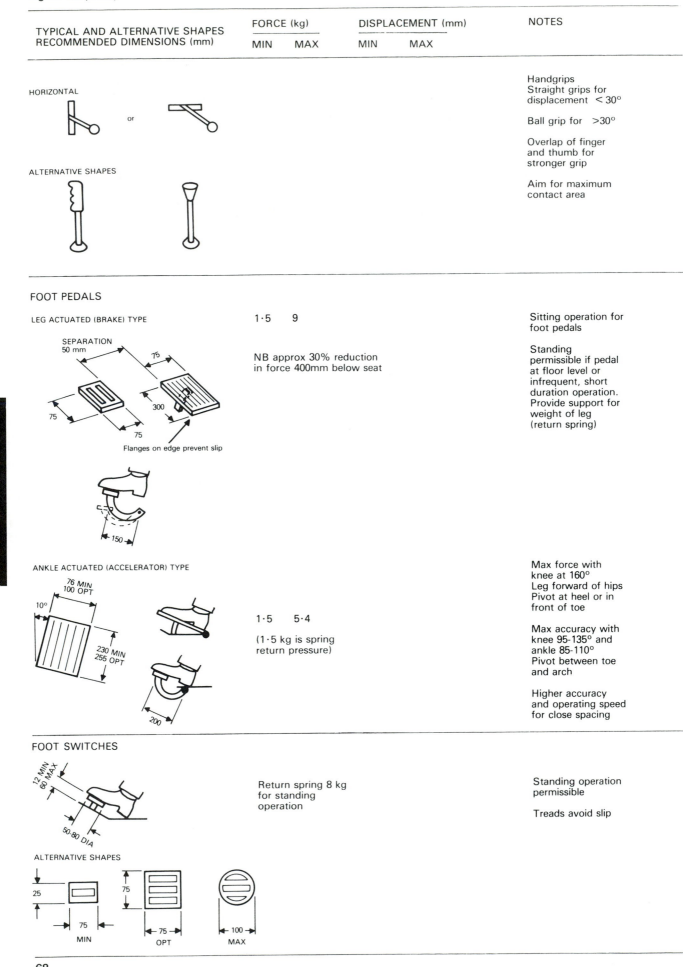

HORIZONTAL

or

ALTERNATIVE SHAPES

Handgrips
Straight grips for
displacement $< 30°$

Ball grip for $> 30°$

Overlap of finger
and thumb for
stronger grip

Aim for maximum
contact area

FOOT PEDALS

LEG ACTUATED (BRAKE) TYPE — 1·5 9

SEPARATION 50 mm
75
300
75
75
Flanges on edge prevent slip
150

NB approx 30% reduction
in force 400mm below seat

Sitting operation for
foot pedals

Standing
permissible if pedal
at floor level or
infrequent, short
duration operation.
Provide support for
weight of leg
(return spring)

ANKLE ACTUATED (ACCELERATOR) TYPE — 1·5 5·4

76 MIN 100 OPT
10°
230 MIN 255 OPT
200

(1·5 kg is spring
return pressure)

Max force with
knee at 160°
Leg forward of hips
Pivot at heel or in
front of toe

Max accuracy with
knee 95-135° and
ankle 85-110°
Pivot between toe
and arch

Higher accuracy
and operating speed
for close spacing

FOOT SWITCHES

12 MIN 60 MAX
50-80 DIA

Return spring 8 kg
for standing
operation

Standing operation
permissible

Treads avoid slip

ALTERNATIVE SHAPES

25
75
MIN

75
75
OPT

100
MAX

CONTROL DESIGN

4.4.1 Further notes

Other types of control or variations on the above may be suitable for particular situations, e.g., slide switches, thumbwheels, roller ball.

The above data are intended only to provide general guidelines for control selection and design, particularly that on forces. The characteristics and abilities of the proposed users and the task requirements must be considered throughout. In some cases control design will be limited by space, commercial availability, etc.

Performance characteristics depend greatly on the strength, fitness, skill and training of the operator, location of the control, direction and distance of movement, duration of the task, etc.

In allowing for strength variations, design for the weakest user (allow for age, sex, fitness, training).

Allow for the duration of the control task. An operator can only exert maximum strength for short periods, with long recovery periods: 5% of maximum strength can be exerted continuously and 20% frequently if rest intervals ten times holding time or more.

4.5 Control – machine relationships

4.5.1 General considerations

Controls are compatible when the machine (or display) moves in an expected direction relative to the control movement, *or* when the controls correspond spatially or functionally to the system being controlled.

Stereotyped movements may be:

   Natural, e.g., control up = system up

   Cultural, e.g., up = OFF (UK) but ON (USA)

In designing control directions consider:
Stereotypes.
Accepted practice.
Consistency.
Standardization.

Some consequences of poor compatibility are:
Reversion to expected direction under stress even if incompatibility has been learnt.
Longer training time.
More errors in the initial response.
Lower speed, precision, reaction time.
Marked effects with age.

Compatibility is especially important where:
Errors are dangerous or costly.
Operations are complex.
The sequence is interrupted.
Training is limited.
There are frequent changes of machine by an operator and non-standardized controls.

Consistency is important, especially where there is no strong stereotype or when laying out groups of controls and displays.

CONTROL DESIGN

69

4.5.2 General rules

The top or left-hand side of the control should move with the display or machine action.

The side of the control nearest the display should move with the display or machine action.

The movement of the control should be proportional to the display or machine movement.

Display movements are according to the line of regard and not the orientation of the body.

Dominance varies with display position relative to the operator, but less so with strong stereotypes.

It is not affected by the preferred hand.

Figure 4.3 Required movement of control for movement of machine.

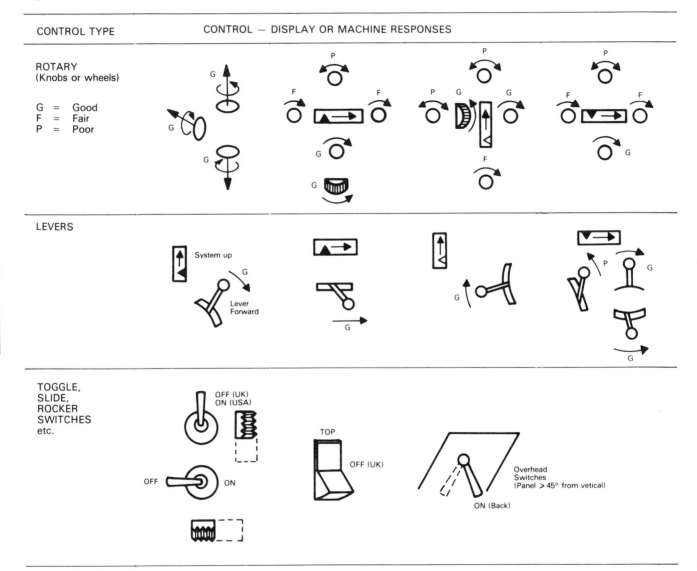

4.5.3 Feedback

Power assistance

For positional control, the machine moves a distance corresponding to the control shift.
For velocity control, the machine keeps moving at a velocity proportional to the movement of the control (this allows flexibility in design and improved performance).

70

4.5.4 'Feel'

Resistance + kinaesthesis (sense of position and movement), e.g., spring loading, to indicate zero or a central position.
NB. High resistance or small displacement masks the 'feel'.

4.5.5 Friction

Static friction

Opposes initial movement, which increases rapidly once it is overcome.

Coulomb friction

Continues to resist movement but does not change with amplitude or speed of displacement. It is useful for preventing accidental operation due to drift, tremor, jolting or weight of limb.

4.5.6 Elastic resistance

This increases with increase in displacement. The level of resistance is the primary cue to the level of control output (it can be nearly zero).
It is often used in combination with displacement to provide a resistance gradient across the displacement range.
Elastic resistance is also called spring loading.

4.5.7 Viscous damping

This increases with an increase in the displacement speed. It is useful for controls which must be displaced at constant speed, and is most effective when used in combination with inertia, which increases with increase in displacement acceleration.

4.5.8 Pressure and amplitude feedback

$$\text{Combined change of sensation} = \frac{\text{Change of resistance}}{\text{Resistance} \times \text{displacement}} = \frac{\triangle F}{F(\triangle D)}$$

(i.e., it is a ratio of proportional change in resistance to change in location).

4.5.9 Sensitivity

To pressure

Less sensitivity to change at lower pressure.

To displacement

$$\text{Displacement ratio } R = \frac{\text{Distance moved by control or display}}{\text{Distance moved by machine}}$$

Small R is for large, rapid movements with low accuracy. Hunting is likely and skilled performance is difficult.
Large R is for fine control, but is slower.
A compromise is needed:

$$R > \frac{\text{Expected operator error}}{\text{Maximum acceptable machine positioning error}}$$

To velocity

The velocity of movement of the display or machine per unit displacement of control.

To gain

This should be balanced against machine lag.

5 Displays and Information

DISPLAYS AND INFORMATION

5.1 Design criteria

Figure 5.1 Flow chart of criteria for the design of displays.

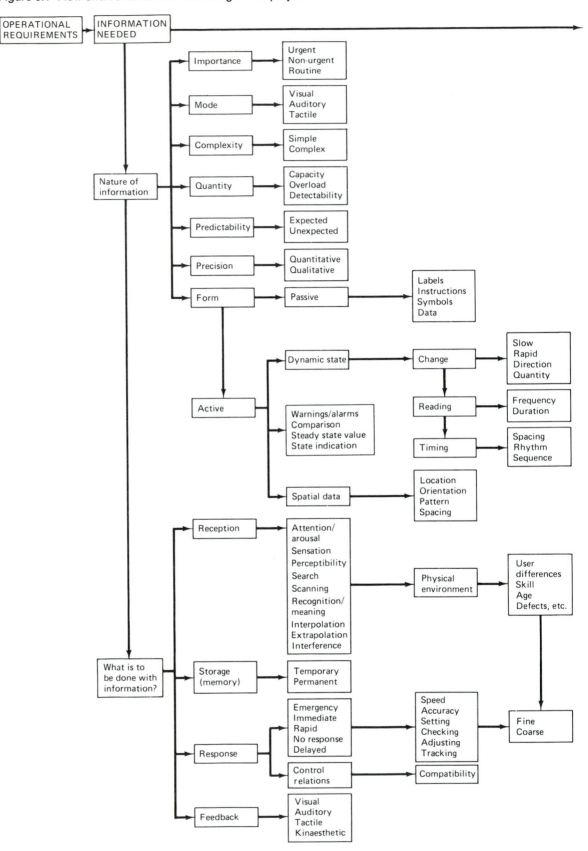

Figure 5.1 (cont.)

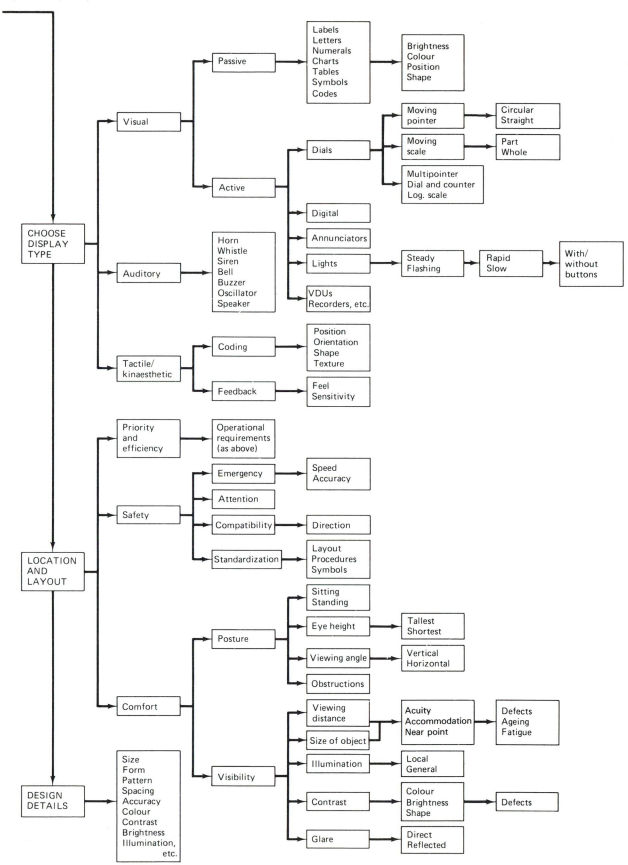

5.2 Choice of display type

5.2.1 General

Choose according to *Design criteria* (p. 74) and the considerations below.
Consider the nature of the information and what is to be done with it, paying particular attention to the difficulties of attending to sensing, interpreting (perception), memorizing and responding to information. Where attention is critical, two cues are better than one.

5.2.2 Lights

This includes illuminated buttons and colour coding.

Warnings/urgent messages: combine with an auditory alarm for gaining attention, flashing or steady.

Discrepancy: flashing light or steady, caution, etc.

Indication: steady.

Confirmation: command received.

Lights are useful for rhythmical information, ease of attention and search, recognition, discrimination and speed of response. With controls they are useful for checking and setting.

5.2.3 Dials

Dials are satisfactory for most information except warnings, and complex and stored information.
They may be used with controls for setting, etc.
Note the various types of dials:

Circular.
Sector.
Straight.
Linear and non-linear (avoid log. scales if possible).
Multi-pointer (avoid except for non-urgent information).
Moving scale (avoid except for non-urgent information).
Mixed dial + digital (for large ranges, high accuracy and direction of change).

Circular dials are better for comparison and rate of change.
Linear dials are better for compatibility and occupy less area for a given scale base.
Interpolation is difficult and errors are more likely with logarithmic scales.

5.2.4 Digital displays

These are used for high accuracy; slow change; quantitive information; frequent, short duration reading; and for setting machines.
They are unsuitable for rapid change or showing direction of change; and for check controlling, check reading or comparison, the position of the dial pointer is a useful visual cue.
Gross reading errors are more likely.

5.2.5 Annunciators

These are indicators or devices displaying a legend illuminated internally on receipt of an initiating signal. They are useful for non-quantitive, especially verbal, information, or for instructions or warnings.

5.2.6 Visual display units

VDUs are necessary for complex information outputs and data inputs.
See *Design criteria* (p. 74) for considerations relating to screen target or signal, visual workspace, etc.

5.2.7 Recording indicators

These are used where storage of information is required.
(Note the limitations of human memory.)

5.2.8 Auditory displays

These are better for warnings and gaining attention, but should not distract or annoy. They should be of a suitable frequency and intensity for the environment.
They are useful for rapid change, rhythmical, spaced, infrequent information, ease of search and discrimination.
The signals are more easily remembered than visual signals, but are more easily interfered with (masked) by the environment.
Examples are horns, whistles, sirens, bells, buzzers, oscillators and speakers.

5.2.9 Other modes

Tactile (touch), kinaesthetic (movement), proprioceptive (position) and olfactory (smell) information may be useful in some systems.

5.2.10 Passive displays

Letters, numerals, charts, tables, symbols, codes and text may be used for warnings, labels, instructions, operating and maintenance procedures, settings, etc.

5.3 Location and layout

See also *Location and layout* (p. 63). Similar conditions apply.

5.3.1 Operational requirements

Arrange for efficient viewing and attention according to priority (frequency, duration, speed, accuracy, sequence) and importance for safety. High priority displays must be in the prime zone of vision (see Table 2.1 and Figure 2.3).

5.3.2 Visibility

Arrange for comfortable vision: a comfortable posture and viewing angle for a range of users in a range of operating postures, seated or standing. Consider the priority of vision as an information source.
A comfortable viewing distance (e.g., 400–700 mm) for the size of object and environmental conditions is necessary.
The viewing angle of surface must be approx. 90° to the line of sight and free from obstruction. Consider short and tall users.
Locate for maximum contrast and minimum glare (see CIBS Code, 1984). Allow for visual and hearing defects and deterioration with age.
The display should be clearly and comfortably visible when the corresponding control is operated.

5.3.3 Search

Do not overcrowd or clutter – search time increases with display density. Standardize the location of displays with respect to each other and the controls, but avoid too much regularity or irregularity.

5.3.4 Compatibility

See *Compatibility* (p. 64).
The movement and layout of displays must be compatible with the controls.

5.3.5 Coding

By position and colour: do not use more than nine colours (see *Colour coding*, p. 85); or by labelling of suitable size and position. Do not obscure.

5.3.6 Grouping

Avoid interference.
Arrange in order of priority according to function.
Displays must be positioned close to the appropriate controls or in a similar pattern.

5.3.7 Sequence

Arrange for a smooth sequence, i.e., in close proximity.
View from left to right and top to bottom.

5.3.8 Checking and comparison

If all pointers in a bank of dials are in the same position for steady state, the odd one out can easily be detected.

5.3.9 Graphic panels

These are useful for indicating flow and spatial systems diagrammatically.
Attention may be more difficult if they are spread over a large area, and if important and unimportant information is mixed.

5.4 Design details

5.4.1 Lights and illuminated buttons

Types

Colour

See *Colour* (p. 82). Choose according to operational requirements, standards or local conventions.
Use filters/lenses according to BS 1376 (1985). Consider classification (A, B or special) and chromaticity.

Brightness

The brightness should command attention under all expected conditions without glare or dazzle.
Vary the size and brightness for extra distinguishability.
The signal brightness of white filters/lenses must be as for yellow.

Warnings

See *Alarms and warnings* (p. 83).
Include a standby lamp or power source in the event of failure (red).

Discrepancy

Any discrepancy between the set position of an indicator and the operating conditions of the equipment causes the light to be illuminated (flashing if required). It may include a mimic diagram.
An associated discrepancy controller may be so arranged that discrepancy is indicated when an operation is selected and cancelled when the operation is carried out (yellow).

Indication

Any specific or non-specific meaning, e.g. start (green), command confirmed (blue, white).

Flashing	Minimize the number. In general restrict to highest priority warnings. Other uses are: request attention and to indicate a discrepancy or change in process. Rapid flashing should be used for higher priority (110 ± 30 flashes per minute) but note that flicker and flicker fusion may occur if the frequency is too high. Use slow flashing for lower priority (20±5 flashes per minute).
Illuminated buttons	Never use for an emergency stop (the lamp may fail).
Test	Provide test circuit so that all lamps can be checked.

Indication

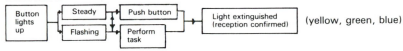

(yellow, green, blue)

Confirmation

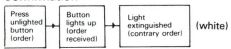

(white)

Double confirmation

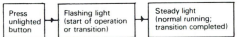

5.4.2 Dial design	Refer to BS 3693 (1992), *Recommendations for Design of Scales and Indexes on Analogue Indicating Instruments.*
Accuracy	1% for 100 interpolated spacings.
Tolerance	Tolerance depends on the system or reading time. It must not be greater than the mechanical tolerance. It must be greater for the commercial than the test system.
Resolution	The resolution is the smallest fraction of the scale range to which the reading is made.
Called interval	This is the physical width or base angle into which scale divisions are divided by eye (interpolated) to design tolerance.
Interpolation	Ten interpolations produces inaccuracy. Five interpolations are better than five marked divisions.
Dimensions	Scale base length $L = 7 \times 10^{-1.3} D$ (in mm). D = viewing distance. From this r can be determined by geometry. For example, the optimum diameter for D = 600 mm is 60–75 mm.
	Allow for the poorest viewing conditions and the poorest eyesight.

DISPLAYS AND INFORMATION

Figure 5.2 The design of dials.

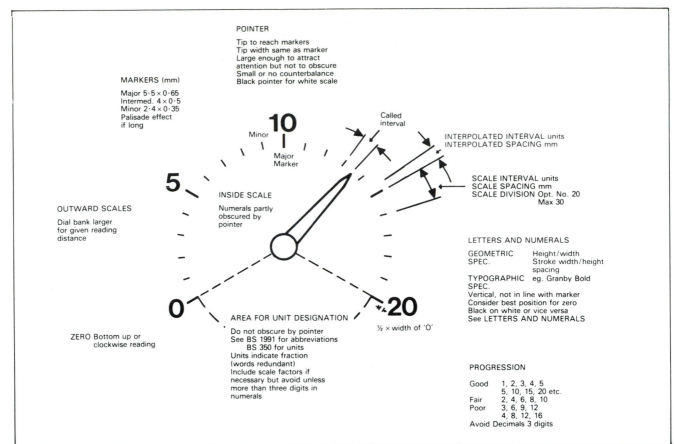

POINTER

Tip to reach markers
Tip width same as marker
Large enough to attract
attention but not to obscure
Small or no counterbalance
Black pointer for white scale

MARKERS (mm)

Major 5·5 × 0·65
Intermed. 4 × 0·5
Minor 2·4 × 0·35
Palisade effect
if long

Called
interval

INTERPOLATED INTERVAL units
INTERPOLATED SPACING mm

Minor

Major
Marker

SCALE INTERVAL units
SCALE SPACING mm
SCALE DIVISION Opt. No. 20
Max 30

INSIDE SCALE

Numerals partly
obscured by
pointer

OUTWARD SCALES

Dial bank larger
for given reading
distance

LETTERS AND NUMERALS

GEOMETRIC   Height/width
SPEC.   Stroke width/height
                spacing
TYPOGRAPHIC eg. Granby Bold
SPEC.
Vertical, not in line with marker
Consider best position for zero
Black on white or vice versa
See LETTERS AND NUMERALS

AREA FOR UNIT DESIGNATION

Do not obscure by pointer
See BS 1991 for abbreviations
       BS 350 for units
Units indicate fraction
(words redundant)
Include scale factors if
necessary but avoid unless
more than three digits in
numerals

½ × width of 'O'

ZERO Bottom up or
clockwise reading

PROGRESSION

Good   1, 2, 3, 4, 5
        5, 10, 15, 20 etc.
Fair   2, 4, 6, 8, 10
Poor   3, 6, 9, 12
        4, 8, 12, 16
Avoid Decimals 3 digits

Refer to BS 3693 parts I and II
Dimensions in mm

SCALE BASE LENGTH

$L = 7 \times 10^{-1.3}D$ mm
D = Viewing Distance
Hence r by geometry
Size: optimum diameter 60 – 75 mm for D = 600 mm
Allow for the poorest viewing
conditions and poorest eyesight.

See calculations in BS 3693

Accuracy	1% for 100 interpolated spacings
Tolerance	Depends on the system/reading time Must not be greater than the mechanical tolerance Must be greater for the commercial than the test system
Resolution	Smallest fraction of scale range of which reading made
Called interval	Physical width or base angle into which scale divisions divided by eye for designed tolerance
Interpolation	10 inaccurate 5 interpolations better than 5 marked divisions

Other dial types

Straight scale

This has some compatibility advantages over a circular dial, e.g., the 12 o'clock effect where downward movement may be incompatible with upward movement of the control.
It occupies less area for a given scale base length then the circular dial.

Horizontal   Vertical

Sector

The markers should be opposite the pointer. Avoid the RH scale.

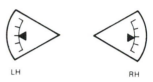

Top   Bottom

Horizontal

LH   RH

Non-linear (logarithmic scale)

Use for a large range. Avoid if possible. Use a digital display for a slow change, and digital plus pointer for a rapid change.
Use depends on accuracy and consequences of error. If it is used, interpolation should be constant.

Multipointer

Use for a large range. Avoid if possible – similar conditions to the log. scale apply (i.e., use digital displays). If used, the pointer and markers must be clearly distinguishable by varying size, thickness and colour.

Moving scale

Avoid except for crude measurements or where the consequences of error are not serious.

Mixed digital plus pointer

Use for large ranges where rapid change, high accuracy and the direction of change are required. Design the dial as above, the digits as below.

DISPLAYS AND INFORMATION

Table 5.1 *The use of lights and colours in displays.*

Colour	Steady indicator lights (clear lens or diffused light)	Rapid flash (110±30 flashes per minute) Significance	Rapid flash (110±30 flashes per minute) On acceptance	Slow flash (20±5 flashes per minute) Significance	Slow flash (20±5 flashes per minute) On acceptance	Push buttons (surface colours as seen in incident white light)
		Flashing lights				
Red	Alarm Warning of potential danger Immediate action	Urgent action to avert danger (plus audible alarm)	Audible alarm (if any) silenced Flash signal changes to steady, extinguished with normal conditions Except for 'Danger' when flashing continues	N/A	N/A	Stop/off
Yellow	Caution Change or impending change	Action required Unwanted change (± audible alarm)	Audible alarm silenced Flashing signal may be extinguished or may continue flashing until action taken *or* may change to be steady signal	Lower priority Unwanted change	Flashing signal changes to steady	Intervention to suppress abnormal conditions or avoid unwanted change
Green	Safety Proceed Command confirmed	N/A	N/A	Changes of state Discrepancy from commanded state Attention required	Flashing signal changes to steady	Start/on General start Start part of machine Close switching device Inching/jogging
Blue	Specific meaning (any not covered by red/yellow/green Command confirmed	N/A	N/A	Changes of state Discrepancy from commanded state Attention required	Flashing signal changes to steady	Any specific meaning, e.g., reset
White	No specific meaning (neutral) Whenever doubt about red, etc. Command confirmed	Non-urgent action required, e.g., change of state or discrepency from commanded state	Audible alarm (if any) silenced. Flashing signal may be extinguished or may continue flashing until action taken *or* may change to a steady signal	Changes of state Discrepancy from commanded state Attention required	Flashing signal changes to steady	No specific meaning Start/on Reset Start/on and stop/off with the same button (better than red or green)
Black/ Grey	—	—	—	—	—	No specific meaning, e.g., inching/jogging (not red) Reset

5.4.3 Digital displays and annunciators

See BS 4099 (1986) Part I (same as IEC 73:1984) *Specification for Colours of Indicator Lights and Push Buttons*; and Part II (1986) *Specification for Flashing Lights, Annunciators and Digital Readouts.*

Digital displays

Use — See *Choice of display type* (p. 76). For accurate, slow changing quantitative information.

Types — Electronic (LEDs) or mechanical counters (on drums). To BS requirements.

Colour — If non-BS colours are used, separate them from the standardized indicator lights and annunciators. Mechanical counters must be black on white or vice versa.

Brightness	There must be adequate visual contrast with the background. Avoid glare: surfaces must be matt or ground.
Numerals	See *Letters* and *Numerals* (p. 86) and the appropriate standards. Spacing between characters: $\geq \frac{1}{2} \times$ height. Stroke and width: $1/6 - 1/10 \times$ height. Height: $\leq 1/360 \times$ reading distance. Read from left to right.

Annunciators

Use	As an internally illuminated legend for conveying information, instructions or warnings.
Colour	See Table 5.1. Preferably use the colour on a dark background.
Illumination	Have adequate brightness but avoid glare or a halo (ground or matt surface). Use rear illumination with two or more parallel lamps for urgent annunciators.
Legend	The legend must not be visible until the lamp is lit. It must be clear and distinct.
Location	Locate according to the importance and the number of operators who need to see it.
Dimensions	*Large size:* Spacing between window centres: vertical 50 mm horizontal 100 mm. Minimum illuminated window area: 40×90 mm. *Small size:* Spacing between window centres: vertical 25 mm horizontal 40 mm. Minimum illuminated window area: 20×33 mm.
Characters	As for digital displays numerals. Height to width ratio approx. 3:2 Use the abbreviations in BS 5775 (1993) (ISO 31 (1992)).

5.4.4 Lights (also auditory), alarms and warnings

Types	
	See *Choice of display type* (p. 76)
Requirements	Ease of attention. Distinguishability. Rapid identification. Rapid response. See *Design criteria* (pp. 74–75).
Visual	See Table 5.1.
Location	Locate in the prime zone of vision for easy attention and in compatible association with the control and control response (combine with the control if possible). See *Location and layout* (pp. 77–78).
Attention	Add an auditory signal where there is high urgency or where several displays must be monitored.

83

Identification	By colour coding (see Table 5.1).
	By labelling.
	Position at a distance from each other.
Flashing lights	Use only for the most important signals or in noisy environments (see *Flashing*, p. 79).
Number of lights	Use as few as possible, e.g., use one main warning light plus a bank of indicators for more detailed information.
Brightness	The light should command attention under all conditions without glare.
	If external lighting is dimmed, dim with a photoelectric cell.

Auditory

Type	See *Auditory displays* (p. 77).
	Choose according to importance, consequences of failure, whether action is required or not, and physical environment (e.g., background noise).
Location	Multi-directional – does not rely on visual attention or facing in particular direction.
	Some can be beamed directionally for greater effect (e.g., horn).
	Include a visual indication in the priority zone of the main panel and display more detailed information on a back-up panel.
	Locate according to hearing and attention requirements plus background noise. Note that sound intensity falls off with the square of the distance from the source.
Attention	Maximum alerting effect is caused by the sudden onset of high intensity sound.
	Variable frequency is more alerting than steady state.
Identification	If several sounds are used, clearly differentiate by quality of sound, frequency and intensity (see *Auditory displays*, p. 77).
Sound level	This must not be painful or damaging (see *Auditory environment*, p. 50).
	It must not be startling, distracting or annoying.
	If intensities are high, include at least two frequencies from the lower end of the spectrum.
	Allow for hearing loss of some users, especially at higher frequencies.
Failure of alarm	Incorporate an emergency power supply for the alarm system if necessary.
	Automatic rectification or standby to be included if the expense is justified.
	Incorporate test procedure.
Rectification/acknowledgement/ silencing	No silencing until the danger over if it is a danger to personnel.
	If it is only a warning, it should be capable of being silenced on acknowledgement.
	If a fault is present, the silencing of an auditory alarm should not cancel a visual alarm until the fault is rectified.
	See *Lights and illuminated buttons* (pp. 78–79).
Reset	The alarm should automatically reset when the fault has been rectified.
	See *Lights and illuminated buttons* (pp. 78–79).

5.4.5 Visual display units

The design of the visual environment for a VDU is influenced by the intensity and contrast of the screen, the rate at which the screen image is electronically replaced (refresh rate), and the need to maintain relatively close eye focus for long periods of time on the screen and working data. General lighting is recommended at 200–300 lux, with local desk lighting (adjustable) for individuals.

Reflections in the screen must be avoided by:
Darker walls, etc. behind user;
Matt screen surface;
Individual tilt and rotate facility for each VDU;
No direct light on screen.

Glare must be avoided by:
No bright surfaces, lights or windows near the line of sight of the screen;
Adjustable window coverings;
Highly diffused room lights.
(See Cakir *et al.*, 1981; Anon., 1983).

5.4.6 Passive displays and visual coding: general

Design criteria

Figure 5.3. Flow chart for the design of symbols for machines.

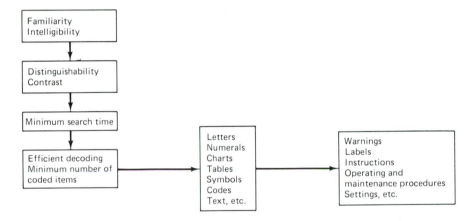

Coding types

Colour

Untrained users (inc. 'colour blind')	Equally discriminable colours	
Red	3R	Select
Orange	9R	those
Yellow	9YR	furthest
Blue	16Y	apart if
Purple	3G	less than
Grey	7BG	9 needed
Buff	9B	
White	9PB	
Black	3R	

Colour is best for rapid search, familiarity, contrast (black on white is best) etc.

A limited number of colours is preferable, ≯ 9, for untrained users.

The search time is much less if the colour of the target is known.

There is the problem of colour vision defects (mainly shortening of the red end of spectrum so that red appears as deep yellow).

Where appropriate add written or auditory information.

Consider the colour specification (CIE chromaticity value).

See *Bibliography* for Standards.

See *Lights and illuminated buttons* (pp. 78–79).

Geometrical shapes	○ △ ▢	Medium search time.
Numerals	1　2　3　4　5 　6　7　8　9　0	Use where possible for quantitative information. Minimize the number of digits (e.g., see *Dials*, p. 76). Use the appropriate dimensions for legibility. Consider the geometrical and typographic specifications. See *Letters, numerals and labels* (p. 87). See *Bibliography* for Standards.
Letters	ABCD　abcd	Lowercase more readable than capitals. Words or sentences are important for quantitive information. Minimize the number of words. Words are more effective if combined with other display types or information. There may be a language problem (an advantage of symbols). There may be a space problem (an advantage of symbols). See considerations and references for *Numerals* (above).
Arrows	→ ↑ ↓ ↻	Consider orientation and length etc.
Graphic symbols	Turning 　　　　MODE 1 Cycle Electric motor　ELEMENT Coarse　VALUES/ accuracy　QUANTITIES	Consider standardization: ISO 3641; BS 3641 (1 to 3); BS 7324 (1990), (ISO 7000). BS 7477 (1991), (IEC 416, ISO 3461.1). Meaningfulness/ambiguity. Grid base. Simplicity. Closure. Continuity. Pictorial quality, line, contrast. Groups of symbols. See *Symbols for machine tool design* (p. 88).
Other codes	✗　Do 　　not　ACTIONS !　Caution　SAFETY	For machine tools consider: Modes; elements. values/quantities; actions; safety and status; and compound symbols;
Size	○　　○　　○	Search times are longer than for other codes. Standardize layouts where possible. Consider compatibility.

DISPLAYS AND INFORMATION

Position/Location		See *Location and layout* (p. 77).

Brightness

○   -○-   -○-

This must not be excessively bright or dull.
It must be brighter for more important displays.
See *Lights* (p. 76).

Combinations

STOP

Colour
Word
Brightness

Better identification can be obtained if codes are combined.

5.4.7 Letters, numerals and labels

Design criteria

Legibility	→	Probability of detection	→	Geometric specification	→	Width/Height Stroke width/Height Black on white White on black
		Size (visual angle)				

Typeface Case Body height Point size 'x' height Cap height Ascenders Descenders etc.	←	Typographic specification	←	MANUFACTURING METHOD Photographic Engraving STANDARDIZATION

Geometric specification

Stroke width → **A** ← Height ↓ width →

Optimum width to height ration = 0·7:1
Stroke width to height ratio = 1:6 for black on white
= 1:8 for white on black
Optimum improved legibility and detection.

Typographic specification

Reinterpret the geometric specification and select a suitable typeface, e.g., Granby Bold numerals, Gill Sans letters (see BS 3693 Part I and BS 2961).
Use an ISO grid for standardization.
Compromise is necessary where space, manufacturing method or usage means that a particular typeface is unsuitable.

Labels

The aim is for easy search and identification in the poorest conditions.
Label controls, displays and other components using letters, numerals, words, symbols, colour and other visual coding, as appropriate.
Use combinations of codes where possible for extra clarity.
Use a minimum number of words while retaining unambiguity.
Avoid redundancy and clutter.
Use standard abbreviations (BS 7477, 1991) and units (BS 350, BS 5775).
Design letters and numerals as indicated, to be of suitable size while allowing for space and typographic restrictions.
Etching or embossing, etc., is more durable than painting.
Standardize the position and direction (horizontal not vertical).

DISPLAYS AND INFORMATION

87

5.4.8 Symbols: general principles

Application

Identification		Place on:
Instruction	→	Equipment and parts Sites and ways Plans, drawings, etc.
Command		
Warning		
Indication		

Design criteria

Unambiguous verbal description Unambiguous symbol		No misinterpretation
Meaningfulness Simplicity	→	Dissimilar and distinguishable from other symbols
Line/contrast Continuity Closure		Clear reproduction, even if small
		Ease of learning and memory if meaning not self evident

Design process
Consider users and potential for understanding symbol(s).
Identify the need for a symbol.
Define the purpose of the symbol – analyse operational requirements.
Consider existing or proposed symbols.
Design, test and modify using the graphic form prescribed in BS 7477 (1991) (ISO 3461) and BS 7324 (ISO 7000).

Combined symbols
Symbols may be combined or grouped.
The meaning of the combination should be unambiguous (use as few symbols as possible).

Graphic form/basic pattern
Refer to BS 7477 (1991) (ISO 3461) and BS 7324 (1990) (ISO 7000).
The basic pattern is a frame of circles, squares, octagons, etc. to aid designers in producing original designs. Forms of suitable line thickness are laid out on this pattern.
Forms should be suitable for economic production by existing techniques, e.g., engraving.
Other drawing aids, e.g., right angles and grids for aligning, are described in above standards.

Orientation of symbols
State explicitly if the meaning depends on orientation.

Colour
Generally black and white are sufficient.
Colour is specified in some cases.

5.4.9 Symbols for machine tool design

Design criteria
See *Symbols: general principles* (above).

Standardization
Refer to BS 7477 (1991), BS 7324 (1990), ISO 7000, ISO 3641, BS 3641 and ISO 369 (1964).
Use ISO and BSI standards as appropriate, except where the latter have been superseded.
Note that some symbols (especially international ones) may be a compromise between various requirements or may be of limited value without training.

DISPLAYS AND INFORMATION

88

Types of symbol

Figure 5.4 Flow chart of the categories of symbols used in machine tool design.

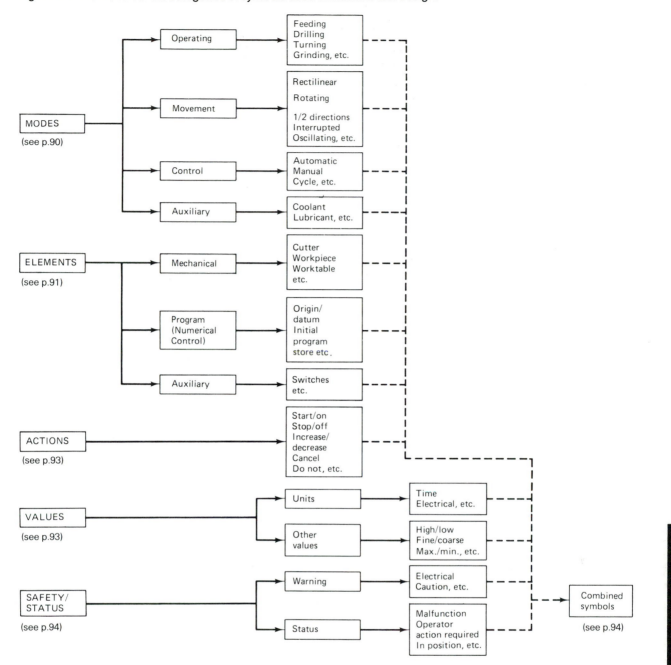

MODES

OPERATING MODES

Bending/folding

Blow

Boring

Broaching chain

Broaching external

Broaching internal

Copying

Drilling

Feed

Grinding

Grinding, cut-off

Grinding, external cylindrical

Grinding, face

Grinding, internal cylindrical

Grinding, plunge out

Honing

Lapping

OPERATING MODES (Cont.)

Milling, climb

Milling, conventional

Milling, vertical

Parting off

Planing

Reaming

Shearing/cutting

Tapping

Threading

Turning

MOVEMENT MODES

Dwell

Feed per stroke

Rectilinear motion

Rectilinear motion in two directions

Rectilinear motion, interrupted (jog)

Rectilinear motion, limited

Rectilinear motion, limited and return

Rectilinear motion oscillating

Rapid traverse

MOVEMENT MODES (Cont.)

Rotation in two directions

Rotation, continuous direction

Rotation, interrupted

Rotation, limited

Rotation, limited and return

Rotation, oscillating

CONTROL MODES

Automatic

Magnetic

Manual

Multi-man operation

One cycle

Step-cycle

AUXILIARY MODES

Balance, dynamic

Balance, static

Blowing unit

DISPLAYS AND INFORMATION

Coolant flood

Coolant mist

Draining

Dressing, crush

Dressing, face

Dressing, form

Dressing, truing

Electrical instruction manual
or diagram

Filling

Gauge size, external

Gauge size, internal

Hydraulic

Lubricant, grease

Lubricant, oil

Maintenance

Material feed

Out of balance

Refer to instruction book

Suction unit

Ventilate blow die

Wall thickness of a preform

Work area illumination

ELEMENTS

MECHANICAL ELEMENTS

Abrasive wheel

Bleed point

Blow mandrel

Blow moulding die

Blow needle

Broach

Broach puller

Broach retriever

Cam

Chain (transmission)

Chuck

Clamp

Clutch

Collet

Conveyor

Cutter (general symbol)

Cutter block

Cutter block
(alternative symbol)

Drive, belt

Drive, gear

Filter

Fixture

DISPLAYS AND INFORMATION

91

Fly-wheel

Foot pedal

Gripper, finger type

Guard

Halfnut

Handwheel

Indicating instrument

Indicator lamp

Injection cylinder

Injection cylinder with plunger

Injection cylinder, with screw

Lever

Main switch

Motor, electric

Motor, hydraulic

Moulded form

Polishing wheel/disc/mop

Pressure cushion (for press)

Pump (general symbol)

Quill

Rectangular work table or slide element

Regulating wheel or feed wheel

Reservoir (tank)

Roller

Rotary brush

Round work table or rotating element

Sanding band

Sanding disc

Sanding drum

Saw, chain

Saw, circular

Saw, linear

Sheer pin

Slide, double facing

Slide, single facing

Slide (press)

Slide (other uses)

Spindle

Strip feed roller

Swarf

Tail stock

Turret, hexagonal

Turret, pentagonal

Turret, square

Work steady

Work head

Workpiece (general symbol)

AUXILIARY ELEMENTS

Continuous material

Float switch

Fuse

Heater

Limit switch

Photo-electric cell

Plug and socket

Preform (e.g. in moulding)

Press tool

Pressure switch

Push button

Selector switch

Solenoid

DISPLAYS AND INFORMATION

AUXILIARY ELEMENTS (Cont.)

Switch

Thermostat

Transformer

ACTIONS

Adjustable

Air eject moulded form

Brake off

Brake on

Disengage

Emergency return of automatic cycle to start

Emergency stop

Engage

In

In action as long as control operated

Lock or tighten

Out

Start and stop with same control

Start/on

Stepless regulation

Stop/off

Trim moulded form

Unlock or untighten

Value, decrease

Value, increase

Work weightpiece

VALUES

QUANTITIES

Amperes

Diameter

Hertz

High

Increment

Low

One revolution

Revolutions

Temperature

Time, hours

Time, minutes

Time, seconds

Volts

Watts

Weight

DISPLAYS AND INFORMATION

WARNINGS

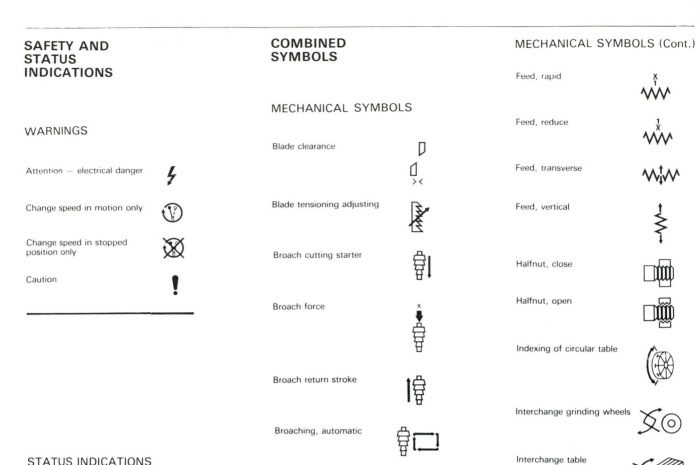

Attention — electrical danger

Change speed in motion only

Change speed in stopped position only

Caution

STATUS INDICATIONS

A.C. supply

D.C. supply

Earth

Malfunction

COMBINED SYMBOLS

MECHANICAL SYMBOLS

Blade clearance

Blade tensioning adjusting

Broach cutting starter

Broach force

Broach return stroke

Broaching, automatic

Broaching, workpiece feeding

Close gripper

Close or lock puller head (broaching machine)

Cutter hold

Cutter release

Direction of spindle rotation

Ejector bar adjustment

Feed per minute
xmm/min

Feed per revolution
xmm/○

Feed, longitudinal

Feed, normal
$\frac{1}{1}$

MECHANICAL SYMBOLS (Cont.)

Feed, rapid
$\frac{x}{1}$

Feed, reduce
$\frac{1}{x}$

Feed, transverse

Feed, vertical

Halfnut, close

Halfnut, open

Indexing of circular table

Interchange grinding wheels

Interchange table

Level, full

Level, low

Load broach into puller

Number of revolutions per minute (spindle speed)
x○/min

Open or unlock retriever (broaching machine)

Peripheral speed of drill
xmm/min

Peripheral speed of milling cutter
xmm/min

Pressure cushion compressed

Pressure cushion gripping

Pump, coolant

Pump, hydraulic system

MECHANICAL SYMBOLS (Cont.)

Pump, lubricant

Shut height (refer to press)

Shuttle table

Slide adjustment

Slide load force (presses)

Speed of boring cut

xmm/min

Speed of planing cut

xmm/min

Speed of turning cut

xmm/min

Stroke adjustment

Tilting table

Tracer, disengage

Tracer engage

Wheel positioning

Work, load

Work, unload

Workpiece conveyor

NUMERICAL CONTROL SYMBOLS

MACHINE TOOL SYMBOLS

Absolute program (co-ordinate dimension words)

Actual position

Axis control in mirror image mode (machine program)

Axis control, normal (machine follows program)

Backward search for beginning of program without machine functions

Backward search for block number without machine functions

Backward search for particular data without machine functions

Backward search for program alignment function without machine functions

Backward tape wind without data read without machine functions

Battery

Beginning of program

Block (basic symbol)

Buffer storage

Cancel, delete (basic symbol)

Compensation or offset (basic symbol)

Co-ordinate basic origin

Data carrier (basic symbol)

Data carrier fault

Data carrier input via an alternative device

Delete store contents

MACHINE TOOL SYMBOLS (Cont.)

Do not (basic symbol)

Editing data in storage

End of program

End of program with automatic rewind to beginning of program without machine functions

Forward block by block read all data with machine functions

Forward continuous read all data without machine functions

Forward continuous read all data with machine functions

Forward continuous read all data with machine functions

Forward search for block number without machine functions

Forward search for particular data without machine functions

Forward search for program alignment function without machine functions

Forward tape wind without data read without machine functions

Grid point (sub-reference position)

In position

Incremental program (incremental dimension words)

Interchange (basic symbol)

Manual data input

Modify, amend, edit (basic symbol)

Optional block skip

Origin/datum (basic symbol)

Positioning accuracy — coarse

Positioning accuracy — fine

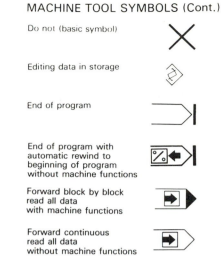

MACHINE TOOL SYMBOLS (Cont.)

Positioning accuracy — normal

Position error (Servo error)

Prewarning storage overflow

Program data error

Program edit

Program from external device

Program storage

Program with machine functions (basic symbols)

Program without machine functions (basic symbol)

Programmed position

Programmed optional stop with machine functions

Programmed stop with machine functions

Read data from store

Reference position

Repositioning

Reset (basic symbol)

Reset store contents

Store (basic symbol)

Storage error

Storage overflow

Subroutine

Subroutine storage

MACHINE TOOL SYMBOLS (Cont.)

Tool diameter compensation (rotating tool)

Tool length compensation (rotating tool)

Tool offset (non-rotating tool)

Tool radius compensation (rotating tool)

Tool tip radius compensation

Write data into store

Zero offset

ADDITIONAL SYMBOLS

Continuous path

Display

Do not

End point of circle

Magazine

Maximum

Minimum

Null

Operator action required
Examine, check

Oriented stop

Per cent %

Planetary milling

Plugboard

Point-to-point

Restart after optional stop

Tool fault,
e.g. missing tool

Tool store

Verified

Groups of symbols

Symbols may be combined to form new compound symbols as follows with new meanings (number of symbols should be minimized to avoid ambiguity).

The new combined symbol may be a: Mode
Element
Value/quantity
Action
Warning or status

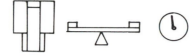 Slide balance pressure

Combined symbols may be used in direct association with controls and are then of the following form (see BS 3641 for examples; see also *Instruction plates*, below).

Figure 5.5 Symbol elements.

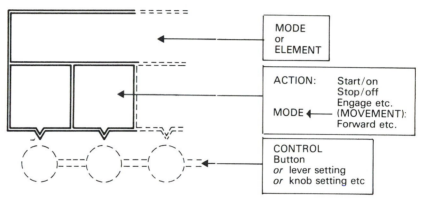

5.4.10 Instruction plates

Aims and criteria

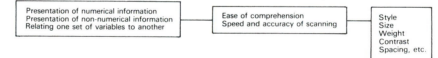

The plate may or may not be associated with controls and displays. See also *Design criteria* (p. 85), *Letters, numerals and labels* (p. 87) and *Symbols: general principles* (p. 88).

Presentation methods

Use the following methods singly, or in appropriate combinations.

Sentences

Avoid prose in bureaucratic style where information must be easily understood.
Keep sentences short, but do not telegraph so that the meaning is lost.
Present lists of simple sentences with appropriate sub-headings where information must be remembered.
Avoid jargon if simpler words will suffice.
Avoid redundant words and information.
Use standard BS 5775 (1993) (ISO 31(1992)) and clear abbreviations.

Logic trees

Use where the user may be uncertain or need help in finding or using information.
Use for trouble shooting.

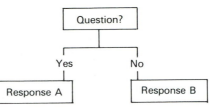

Symbols

Use standard symbols where possible. They are useful where language is a problem, where space is limited and where rapid comprehension is needed (see *Types of symbol*, pp. 89–96).
They are particularly important for use in tables of control settings.
Locate the symbols next to the appropriate controls (see *Combined symbols*, p. 94, and *Groups of symbols*, p. 97).

Illustration

Use where possible for clarity and to avoid technical jargon.

Tables

Tables are preferable to other methods if the user knows what to look up (faster and fewer errors).
They are used for relating one set of variables to another, e.g., machine settings.
They give a systematic arrangement of numerical and non-numerical information.
Full, direct (explicit) presentation is preferred.
This can be linear (list) or 2-dimensional (matrix).
Scanning only is required.
Avoid inferences or combinations of information from various parts of the table (implicit).

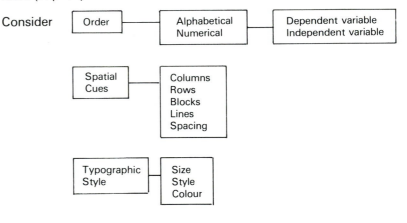

See also *Aims and criteria* (p. 97).

5.4.12 Tabular display design

Order

Position the machine variable to the right of the operating variable (read left to right).

Arrange the operating variable in numerical order to decrease downwards in column.

Use space/lines to facilitate horizontal/vertical scanning.

Spatial cues

Use spaces to group into threes or fives down column.

Groups of 10 are unsatisfactory but better than no group.

Minimize the space between paired items in adjacent columns (minimize the alignment error).

Vertical rules are to separate only unrelated variables in adjacent columns.

Omit redundant abbreviations within the table, e.g., units.

Omit landmarks or extra information within columns, especially where larger/bolder type can be used.

Typography

8–12 point 'x' height is recommended.

Dependent and independent variables must be the same typeface but differently weighted, e.g., the independent variable must be heavier in colour (see *Letters, numerals and labels*, p. 87).

Location of instruction plates

Consider design criteria in the location and layout of displays.

Locate the instruction plate close to the appropriate controls, setting or other operation so that instructions and work, controls etc. can be viewed/operated simultaneously (avoid the need to memorize information or adopt an uncomfortable posture to see the instructions, etc.).

Instructions should not be obscured by materials, covers, components, dirt, etc.

Technical manuals

Similar conditions apply to the design of technical manuals for operation or maintenance.

See BS 4884 Parts I and II, *Specification for Technical Manuals*, or any guide for writing manuals or instructions.

6 *Maintainability*

MAINTAINABILITY

6.1 Introduction

Although much work is being conducted to improve equipment reliability in attempts to increase machine availability, very little work is being directed to maintainability. This is surprising as the cumulative downtime associated with routine and breakdown maintenance can be considerable. Until engineers can design a machine which requires no maintenance, its maintainability features will be important factors in determining both machine operational time and the safety of the maintenance crews.

The benefits of 'maintenance friendly' machinery can be less downtime, reduced health and safety problems and improvements in machine reliability. A machine which is difficult to maintain will probably receive less attention than one that is easy to maintain.

It is often said that optimizing operability of a machine is best achieved by applying ergonomics at the design stage. Although still desirable, designing for maintainability can usually also be satisfactorily addressed on current equipment with retro-fit modifications being made as part of routine refurbishment. The guidelines presented in this section therefore have relevance to both new and existing plant and equipment.

6.2 Design criteria

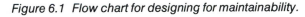

Figure 6.1 Flow chart for designing for maintainability.

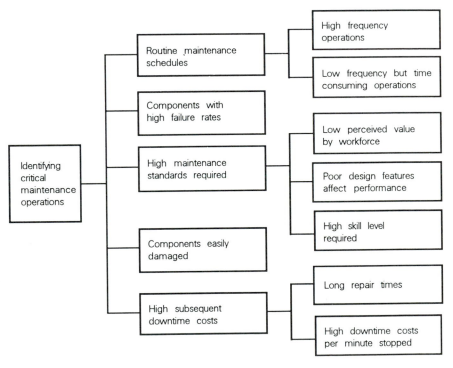

(a)

Figure 5.1 (cont.)

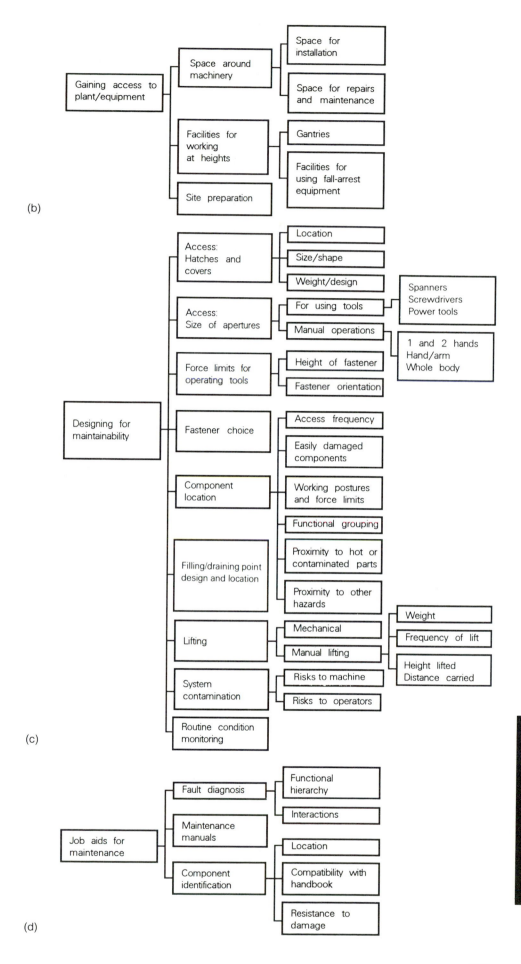

(b)

(c)

(d)

103

6.3 Identification of critical maintenance operations

6.3.1 Maintenance schedules

In many situations it is unlikely that a designer can fully optimize the maintainability features associated with all the maintenance operations on a machine or plant. Space restrictions, for example, often require a trade-off between creating extra space for some tasks at the expense of less space for others. It is therefore useful to identify the more important maintenance tasks. Depending on specific needs, the designer may decide to focus on reduced downtimes and costs, or enhanced equipment reliability. The more important factors are listed for each.

Reducing cumulative costs

The *cumulative time costs* of performing even relatively simple routine maintenance tasks can be surprisingly large when the operation is performed regularly – say each shift. Similarly, although some tasks are performed relatively infrequently they can take excessive times. Components with known high wear rates should also be considered as a priority for designing for maintainability.

The overall costs for performing maintenance tasks may not just reflect the downtime. Some downtime will be expensive, e.g. if it causes sudden lost production. Other downtime will simply reflect overhead costs associated with the maintenance crew time costs. The full cost to the organization needs to be taken into consideration when assessing the importance of each task.

The maintenance tasks should therefore be ranked in terms of their cumulative expected downtimes and costs, with the highest receiving priority attention.

Enhancing equipment reliability

Some maintenance operations require a higher standard of skill/knowledge and behaviour than others. Of special importance are tasks which craftsmen often consider as of little importance to the 'health' of the machine. Unless these tasks are easy to perform they have an increased likelihood of being neglected. Poor maintainability features (among other factors) influence the reliability with which such maintenance operations are likely to be conducted.

Components which are easily damaged need careful attention to minimize any maintenance difficulties.

6.4 Design details

6.4.1 Access

Hatches and covers

For safety and ease of use, access covers should incorporate the following features – see also Chapter 2, *Workspace Design*.

Wherever possible, hinged covers and quick release fasteners should be used. Hinged covers should open to provide sufficient space for all tasks and the hinges should preferably be positioned at the bottom.

Suitable latches or stays should be provided to prevent heavy hinged covers swinging open or closed due to machine movement, inclination of the machine or accidental operation.

Hatches and covers should preferably be designed for use by a single operator. The maximum weights of covers will be dictated by restrictions to working posture – i.e. the closer to the body the more a person can lift. Some guidelines for safe lifts are given in Chapter 2.

Heavy hatches should be capable of being supported on a ledge to avoid the need for precise control of heavy/bulky objects. The ledge should ensure hatches align with any fixing holes.

Wherever possible, cover fasteners should be of the same size and type to minimize the numbers of tools needing to be carried. Retaining chains or other devices may be needed to prevent such fasteners becoming lost.

Appropriately designed and located handles or handholds should be provided (see Section 2.4.3). They may need to be recessed or hinged if they could cause a safety hazard.

Handles should be located relative to the C of G to prevent the object swinging or tilting when lifted, and take into account any restrictions in postures, such as limited headroom. Handles should have no sharp edges and exceed the following minimum dimensions:

Weights up to 6 kg –   6 mm (1/4") diameter
Weights up to 9 kg – 12 mm (1/2") diameter
Weights over 9 kg – 19 mm (3/4") diameter

Handle length   – 145 mm minimum
Handle clearance – 50 mm minimum

Aperture location

Wherever possible, access apertures should be located so that only one cover needs to be removed to replace or service any single component. Apertures should be located on the same side of the machine as any related displays, controls, test points, etc.

Access apertures should not be located in positions where they are likely to create a health and safety risk for maintenance personnel.

Access covers may need to be designed and located to minimize the damage risk resulting from the ingress of dirt/debris. For some machines, it may be necessary to avoid hatches/apertures located on the top surfaces or other locations where they can collect debris, dirt or water which can contaminate a machine or require lengthy cleaning operations. Access apertures should be well located and large enough to eliminate the need to remove non-affected components in order to reach items requiring routine attention.

Aperture dimensions

See also Section 2.4.3.

The size and shape of an aperture should be that which allows the easiest grasping and passage of components, provides sufficient clearance for using tools and enables both the tools and component to be seen while it is being maintained.

Visual access should be provided if people could encounter a hazard inside the aperture.

The exact size of aperture will depend on:

1. The height off the floor.
2. The depth of reach.

Once these are known, and the designer has also determined whether full body access, hand/arm, or hand-only access is required, then the designer can determine:

3. The aperture width and height.

Figures 6.2 to 6.4 should be referred to for detailed recommendations. These dimensions may need to be made larger where there is a need to grasp and carry large components through the aperture. In these instances the data shown in Figure 6.5 should be used.

MAINTAINABILITY

Figure 6.2 Access requirements for two-handed maintenance tasks.

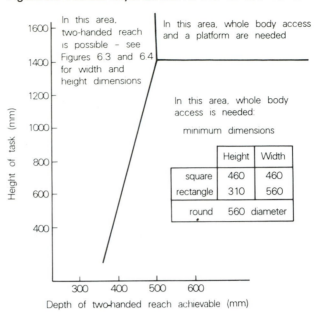

In this area, two-handed reach is possible – see Figures 6.3 and 6.4 for width and height dimensions

In this area, whole body access and a platform are needed

In this area, whole body access is needed:

minimum dimensions

	Height	Width
square	460	460
rectangle	310	560
round	560 diameter	

Height of task (mm)

Depth of two-handed reach achievable (mm)

The terms used in Figures 6.2 to 6.4 are shown below:

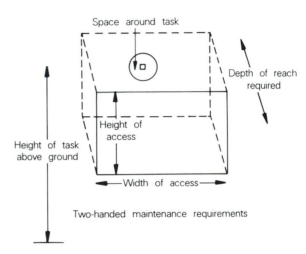

Space around task

Depth of reach required

Height of access

Height of task above ground

Width of access

Two-handed maintenance requirements

Figure 6.3 Width of apertures.

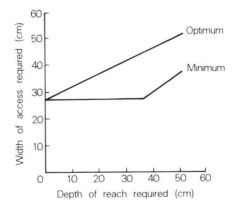

Optimum

Minimum

Width of access required (cm)

Depth of reach required (cm)

Figure 6.4 Height of apertures.

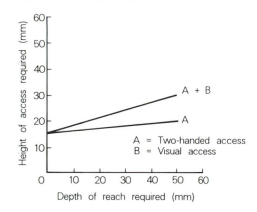

Figure 6.5 Aperture dimensions for access to large components.

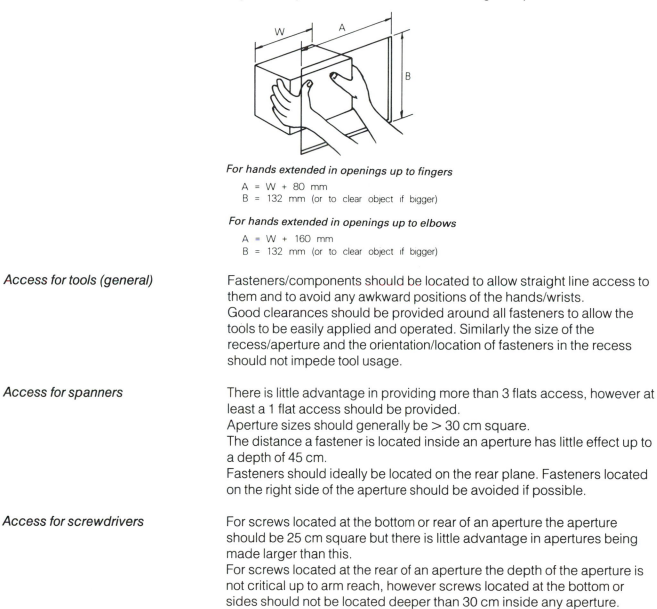

For hands extended in openings up to fingers

A = W + 80 mm
B = 132 mm (or to clear object if bigger)

For hands extended in openings up to elbows

A = W + 160 mm
B = 132 mm (or to clear object if bigger)

Access for tools (general)

Fasteners/components should be located to allow straight line access to them and to avoid any awkward positions of the hands/wrists.
Good clearances should be provided around all fasteners to allow the tools to be easily applied and operated. Similarly the size of the recess/aperture and the orientation/location of fasteners in the recess should not impede tool usage.

Access for spanners

There is little advantage in providing more than 3 flats access, however at least a 1 flat access should be provided.
Aperture sizes should generally be > 30 cm square.
The distance a fastener is located inside an aperture has little effect up to a depth of 45 cm.
Fasteners should ideally be located on the rear plane. Fasteners located on the right side of the aperture should be avoided if possible.

Access for screwdrivers

For screws located at the bottom or rear of an aperture the aperture should be 25 cm square but there is little advantage in apertures being made larger than this.
For screws located at the rear of an aperture the depth of the aperture is not critical up to arm reach, however screws located at the bottom or sides should not be located deeper than 30 cm inside any aperture.

Force limits

Heavy duty fasteners should not require torque levels which exceed safe limits. However, the torque levels recommended often exceed the abilities of all or some of the workforce. In such a case component failure can result if the fastener vibrates loose. Injury can also result where the worker attempts to achieve very high force levels on tools by unapproved methods. Torque levels should be based on the force capacity values of Figure 6.6.

The forces people can exert on spanners have been found to be highly dependent on:

1. The height of the fastener off the floor.
2. The direction of application – both fore/aft and up/down.
3. A facility to brace themselves to apply more force.

Figure 6.6 Target force limits for handtools.

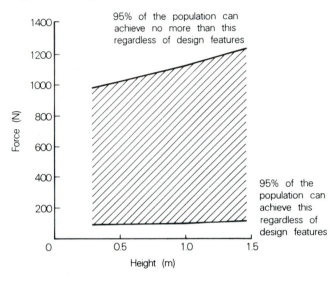

The approximate relationship between fastener height, direction of force application and the effects of ability to brace against a surface are shown in Figure 6.7.

This shows that when located low down, fasteners should ideally be located on the horizontal plane and the ability to brace oneself is very important. At high heights the fasteners should ideally be located on the vertical plane.

It is more usual for the torque required to be known rather than the force required to apply on the end of a typical spanner. The upper and lower limits of ability have been calculated for conventional spanners for fastener sizes from 5 to 50 mm. These are shown in Table 6.1 for three heights of fastener.

Figure 6.7   Relative influence of design features at 0.3, 1.0 and 1.45 m height.

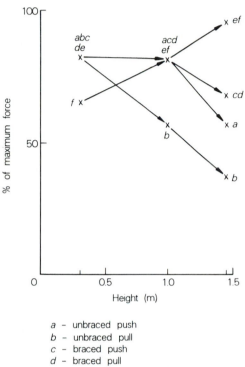

a – unbraced push
b – unbraced pull
c – braced push
d – braced pull
e – upward movement of spanner
f – downward movement of spanner

Table 6.1 Table for converting critical forces into torques for nut sizes ranging from 5 mm to 50 mm.

Nut size (mm)	Spanner length (mm) [1]	0.3 m height		1.0 m height		1.45 m height	
		Lower limit (Nm)	Upper limit (Nm)	Lower limit (Nm)	Upper limit (Nm)	Lower limit (Nm)	Upper limit (Nm) [2]
5	70	6.3	69.3	7.2	79.1	7.9	86.7
7	88	7.3	87.1	9.1	99.4	9.9	108.9
8	97	8.7	96.0	10.0	109.6	10.9	120.1
9	106	9.5	104.9	10.9	119.8	11.9	131.2
10	115	10.3	113.8	11.8	129.0	13.0	142.4
11	124	11.2	122.7	12.7	140.2	14.0	153.5
13	142	12.8	140.5	14.6	160.2	16.0	175.8
14	151	13.6	149.5	15.5	170.6	17.0	186.9
15	160	14.4	158.4	16.5	180.8	18.1	198.0
17	178	16.0	176.2	18.3	201.1	20.1	220.4
18	187	16.8	185.1	19.2	211.3	21.1	231.5
19	196	17.6	194.0	20.2	221.5	22.1	242.6
21*	214	19.3	211.8	22.0	241.8	24.2	264.9
22*	223	20.1	220.7	23.0	252.0	25.2	276.1
23*	232	20.8	229.7	23.8	262.2	26.2	287.2
24*	241	21.7	238.6	24.8	272.3	27.2	298.4
26*	259	23.3	256.4	26.7	292.6	29.2	320.6
27*	268	24.1	265.3	27.6	302.8	30.2	331.8
28*	277	24.9	274.2	28.5	303.0	31.3	342.9
30*	295	26.5	292.0	30.4	333.3	33.3	365.2
32*	313	28.2	309.9	32.2	353.6	35.3	387.5
36	349	31.4	345.5	35.9	394.3	39.4	432.0
41	394	35.5	390.0	40.6	445.2	44.5	487.8
46	439	39.5	434.6	45.2	496.1	49.6	543.5
50	475	42.8	470.2	48.9	536.7	53.7	588.0
10* adj	200	18.0	198.0	20.6	226.2	22.6	247.6

[1] The length of spanners, end to end, is found to be roughly 10 × (size of *larger* jaw in mm) + 25 mm. The effective length of the lever is therefore taken to be 9 × (size of larger jaw in mm) + 25 mm.

[2] Torques which are between the upper and lower limits fall within the shaded area of Figure 6.6.

* These nut sizes can be accommodated by a 10-inch adjustable spanner, therefore the worst case is given by the torque values for this tool at the bottom of the table.

In cases where the human force capabilities are likely to be exceeded, sufficient access needs to be provided for using powered handtools. Power take-off points should be easy to use and well located. Consideration should also be given to the use of hydraulic stud tensioners. Devices such as 'rotabolts' can be used to make torque level checks easier. Once torqued correctly the deformation inside the bolt 'grips' a central shaft. If this shaft can be freely turned, the torque must have slackened and the bolt requires tightening. Such devices eliminated the need for physical checks using torque wrenches.

Fastener choice

On exposed surfaces, fasteners should be chosen which do not suffer from compacted dirt (e.g. socket head bolts), and either recessed or protected from damage.

Component location

All service points, items which require adjustment and components which are likely to need replacing should be clearly visible and easily accessible. For example, fuses should be grouped together where they can be easily checked and replaced without the need to remove other components.

Items requiring only visual inspections should be located so that they can be easily seen without removing covers.

Components which need the most frequent attention should be located in the most accessible locations.

Components regularly serviced should be located away from hot surfaces.

Components should not be located so that access to them is impeded by other non-affected components.

Components should be located to minimize the risk of injury during inspection or removal.

Easily damaged components should not be positioned where frequent or heavy maintenance work is necessary.

Functionally similar components requiring routine servicing should be grouped together.

All test points should be easily accessible and grouped together near any controls which are used as part of the testing operations.

Where good access cannot be achieved, or where components are heavy, consideration should be given to providing pull-out drawers or racks. For heavy components these should have end stops which can be easily released.

Filling and draining

All grease points, filling and draining points should be easily accessible from safe locations. For difficult to reach locations, guide tubes and extended fittings may be needed to avoid dismantling equipment.

Where practical, all filling points requiring the same fluid should be of similar design and, in order to minimize errors, not interchangeable with those used for other fluids.

Labels should be located near fluid filling caps giving the capacity, type of fluid, and any level restrictions.

All filling points should be located to minimize the damage risk resulting from the ingress of debris and to minimize any necessary cleaning before maintenance operations can be performed.

Hand-pumping oil into machinery can be laborious. A powered oil filling facility should therefore be provided on those machines requiring oil to be pumped in if the temptation to pour unfiltered oil by unauthorized methods is to be eliminated.

Drain plugs should be positioned so they do not require a person to crawl under the machine to remove/replace them.

A facility should be provided to allow fluids being drained to be collected in a container as opposed to draining onto the floor or onto sensitive equipment.

Filling points should be located to minimize the consequences of any spillage, i.e. away from easily damaged equipment.

Lifting

The precise nature of the task (e.g. body posture, characteristics of the lift) and the environmental influences (restricted headroom, floor conditions, etc.) should be taken into account when determining the maximum lifting limits – see Section 2.4.4.

Consideration should be given to the design and provision of mechanized lifting points for a machine as an entity and in major component form.

MAINTAINABILITY

Lifting point provisions should take into account the likely lifting equipment available on site and any physical restrictions (e.g. limited headroom).

Handling point locations should take into account the full range of manoeuvres likely (e.g. turning over or dragging) and not simply the straight lift.

Handling points should not be located close to any easily damaged components.

The weight of machine components should be clearly marked. Where a lift requires the use of different lengths of slings these should be specified on permanently attached labels.

System contamination

All specified methods of condition monitoring, routine maintenance operations, and fault diagnostic checks should be able to be undertaken without risk of introducing dirt, etc., into the machine hydraulic/lubrication systems.

Non-invasive methods of checking and testing should be fully exploited, particularly with regard to servicing the hydraulics and lubrication.

In environments where dirt/debris will accumulate on a machine, access through the top of the machine should be avoided.

Use of transparent covers should be considered where only visual inspections are normally required.

All replacement components which could have a contaminating effect on a machine, or which could become damaged/contaminated themselves should be supplied in protective packaging.

The requirements for special cleaning methods/materials should be considered for their practicality in use in the field.

Components which could become cross-contaminated during servicing should be separated (e.g. separating hydraulics from electrical units).

Plug-in electronic cards should be orientated so that dirt cannot fall into the electrical 'sockets' when cards are removed.

Routine condition monitoring

Sample points should be grouped together in a way which reflects the layout of the machine components in order to minimize confusion.

All sampling and test points should be located away from any safety hazards and be highly accessible.

Test points should be located close to any controls or displays which are used in the test process.

Test points should be labelled with the tolerance range and/or limits indications to avoid the need for operators to remember critical values.

Sample points should be located and designed to eliminate any risk of contaminating the samples during the tests (e.g. dirt falling into a hydraulic tank when the filler plug is removed).

Gearboxes and hydraulic tanks should be fitted with sight glasses to check levels, as opposed to level plugs, and should show the full range between high and low levels.

Warning devices should be considered to warn operators of the approach of a machine cut-out situation or failure so that avoiding actions can be taken (e.g. slowing machine down).

6.5 Job aids for maintenance

6.5.1 Fault diagnosis

Faults in any of the major machine modules should be capable of being quickly detected and traced by staff with normal training and skills.

Diagnostic facilities (paper or computer based) should be provided to readily show the functional module with a fault. These should be arranged hierarchically.

Machine diagnostics should be usable when the main power circuits are inoperable.

6.5.2 Maintenance manuals

A comprehensive maintenance schedule should be provided.
The schedule should accurately reflect the real needs of the machine and not be over-demanding in the hope that 'at least some of it may get done!'
Schedules which are over demanding or are perceived as unrealistic often lead to the gradual disuse of wide parts of the recommended schedules.
The schedules should take into consideration the working environment and duty cycle of machines in different situations.
Manuals should clearly draw attention to any potential hazards involved in any operations and the safe methods of working and use of safety equipment.
Maintenance manuals must be kept fully up to date.

6.5.3 Component identification and labelling

See also Chapter 5.
Terminology used for component identification in manuals, labelling of parts, diagnostic materials and any other source of information should be consistent.
The length of serial number should not be more than 17 characters and may consist of numerals and letters. Grouping them on labels into small groups of two to four enhances ease of memory and reduces the possibility of errors in communication.
Maintenance tasks should be assessed for those which could benefit from instruction labelling or warnings.

Procedures should be presented on adjacent instruction plates when:

1. It is not possible to refer to an instruction manual.
2. The task consists of a series of potentially confusing steps.
3. The required response time prevents the use of manuals.

Warning labels should be prominently displayed and preferably contain the following information:

1. Why the dangerous condition exists.
2. Parts of the equipment to avoid.
3. Behaviours to avoid.
4. Sequences of actions to avoid the risk.
5. Where to refer to further information if required.

Such labels should be located on permanent features of the machine which are unlikely to be replaced or damaged.
Where a delicate or dangerous component has to be reached which can not be viewed directly, a label or diagram of the layout should be provided near the access opening.
All identification labels should be robust in design and located to minimize any damage or coverage by debris, to be easily read and not to be obscured by the operator's hand during the work.
Test and service points should be labelled with any reference data which may be required by the maintenance operator.

7   Designing for Safety

7.1 Context and structure

Ergonomic design for safety requires taking into account all possibilities of use of the equipment, including misuse, both when the equipment is new and when wear may have reduced its reliability. Where similar equipment already exists, failure and damage records may be investigated, as well as accident and injury records. Human failures should not be dismissed as 'human error', but considered as possibly reasonable actions of the individuals at the time, where their understanding of the situation might have been falsified by stress, lack of training, overload or social or operational difficulties. Only where no other explanation can be offered should irrationality be assumed, and even then on a provisional basis pending more information.

Safety design requires input from at least designers, constructors, commissioning staff, users, maintenance people and management. All of these influence the way the product is used, and hence its level of safety. The 'design for safety' working group should be available throughout the period of design, to consider major changes to specification, or other matters which may affect the subsequent safety of the product. Summarized flow chart for safety-by-design procedure (see Figure 7.1):

1. Form safety working group with appropriate members, see above.
2. Acquire information on histories, previous failures, accidents or injuries and familiarize the group with the particular site or conditions.
3. Identify system hazards, quantify where possible.
4. Identify job hazards, analyse separately for each phase of construction, commissioning, operating, maintaining, decommissioning and disposal.
5. Prioritize identified hazards and assess for technical solutions and costs. Make design decisions.
6. Implement design solutions.

Figure 7.1 Flow chart for safety-by-design procedure.

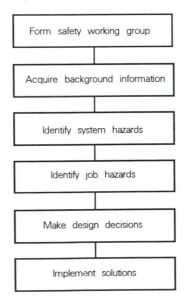

7.2 Acquire background information

7.2.1 Relevant safety standards

These should extend beyond the technical standards to embrace standards for operation and those for the environment. Liability devolves onto the supplier where injury or damage arises from events attributable to design limitations which were technically unnecessary. Where ergonomics standards specify operational or environmental requirements, the omission of which would introduce stress on operators, errors and accidents can be attributable to such stresses.

7.2.2 Previous systems performances

Maintenance and operations records for previous or similar systems should be used to assess distribution and types of likely failures or errors. Where third parties have been used for construction or commissioning of previous systems, their records too should be reviewed.

Associated systems influences

Identify and specify all systems which may have an influence, or introduce a hazard to the operation etc. of the target system. For example,
moving vehicles colliding with the system, failure of neighbouring devices or extreme weather conditions are possible changes which could cause the target system to malfunction.

Other relevant sources

Other sources for information and experience of the safety of similar systems are consultants, insurance companies, government safety inspection bodies. Research organizations of public or private laboratories, including universities, are also potential sources which can be approached. A comprehensive range of data at this stage gives a wider basis for hazard assessment later in the analysis.

7.2.3 Familiarization

The work group should have a common understanding of the system's process, operation and site to ensure a common focus for discussion amongst those with differing skills who form the group. For this a 'walk-through' of all activities of the system, based usually on a task analysis, (see Section 1.2.4), will draw attention to potential operating difficulties. This is conducted, usually in real time, by a plant operator, construction worker or other relevant person, describing in detail the sequence for each activity.

7.2.4 Task analysis

The information collected using task analysis (see Section 1.2.4) will provide much of the information needed to identify job hazards.

7.3 Identify system hazards

The object of this step is to identify the hazards to people passing through or working in close proximity to the system being designed. Hazards to system users are identified at the next stage, Section 7.4.

1. Break the system into sub-systems, e.g. from component or sub-assembly schedules; give location in proposed operation site.
2. List all physical inputs and outputs to/from the sub-systems which, if not controlled could give rise to hazard, e.g. water, steam or chemicals, vehicles.
3. List all neighbouring equipment which might interact to cause a hazard, e.g., piping and ducting, emergency equipment, access routes.
4. Identify and list all potential hazards, using a checklist such as that of Figure 7.2. Be specific and quantify as far as possible.

117

Figure 7.2 System hazard trigger checklist.

Potentional for:
[ ] Objects/People falling
[ ] Explosion
[ ] Flooding
[ ] Fire
[ ] Corrosion/Contamination
[ ] Leakage

Hazards due to:
[ ] Walking surfaces
[ ] Passageways/Aisles

[ ] Moving equipment/parts
[ ] Access/Egress
[ ] Roadways
[ ] Climate/Weather
[ ] Temperate
[ ] Lighting
[ ] Vibration
[ ] Noise
[ ] Electrical source
[ ] Pressurized systems
[ ] Others

7.4 Identify job hazards

The objective of this step is to identify the job- or task-related hazards which, if not controlled, can lead to personal injuries, accidents, physical discomfort, fatigue and mental stress for those working with the system. It assesses the potential hazards from technical factors, including their interactions, and then how these may be affected by personal experiences to create or aggravate a hazard, see Figure 7.3.

Figure 7.3 Hazard identification process.

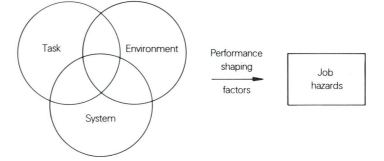

1. Break into appropriate sub-systems, or use those from Section 7.3.
2. Analyse separately, jobs performed during each appropriate phase of the system's life, e.g. construction, operation, maintenance and decommissioning.
3. List all work tasks/activities for each sub-system; use task analysis procedures, see Section 1.2.4.
4. List all tools and equipment, including protective equipment and job aids required to perform the tasks.
5. Describe all task requirements, physical and mental, using the list in Figure 7.4. The information from the task analysis procedure (Section 1.2.4.) will also be helpful here.

Figure 7.4 Task requirements checklist.

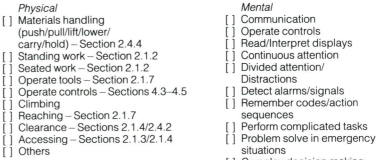

Physical
[ ] Materials handling (push/pull/lift/lower/ carry/hold) – Section 2.4.4
[ ] Standing work – Section 2.1.2
[ ] Seated work – Section 2.1.2
[ ] Operate tools – Section 2.1.7
[ ] Operate controls – Sections 4.3–4.5
[ ] Climbing
[ ] Reaching – Section 2.1.7
[ ] Clearance – Sections 2.1.4/2.4.2
[ ] Accessing – Sections 2.1.3/2.1.4
[ ] Others

Mental
[ ] Communication
[ ] Operate controls
[ ] Read/Interpret displays
[ ] Continuous attention
[ ] Divided attention/ Distractions
[ ] Detect alarms/signals
[ ] Remember codes/action sequences
[ ] Perform complicated tasks
[ ] Problem solve in emergency situations
[ ] Complex decision making
[ ] Others

6. Having identified the work tasks and task requirements, note the presence of any environment factors (quantify where possible) which may affect the worker's ability to perform the tasks safely, see list in Figure 7.5.

7. Use the checklist of performance shaping factors (factors relating to the worker's environment or task which affect performance) to determine any other factors which may affect the worker's ability to carry out the tasks required safely, see list in Figure 7.5.

Figure 7.5 Factors affecting safe task performance.

Environmental factors

[ ] Height	[ ] Radiation
[ ] Noise – Section 3.2	[ ] Vibration
[ ] Temperature – Section 3.3	[ ] Distractions
[ ] Light – Section 3.4	[ ] High traffic areas
[ ] Dust	[ ] Others
[ ] Confined space – Section 2,4	

Performance shaping factors

[ ] Training	[ ] Feedback
[ ] Experience	[ ] Monotonous work
[ ] Health	[ ] Motivation/Attitudes
[ ] Environment	[ ] Emotional state
[ ] Work hours/breaks	[ ] Social factors
[ ] Shift rotation	[ ] Mental overload/underload
[ ] Time pressures	[ ] Human–machine interface
	[ ] Others

Figure 7.6 Job hazards checklist.

Job hazards

Potential for:	
[ ] Musculoskeltal injury	[ ] Inhalation/absorption/ ingestion of chemicals
– Back	[ ] Cut/Abrasion
– Hand/Wrist	[ ] Burn
– Arm/Shoulder	[ ] Asphyxia
– Neck	[ ] Visual impairment
– Other	[ ] Hearing impairment
[ ] Fatigue/Over-exertion	[ ] Radiation exposure
[ ] Slip/Trip	– Ionizing
[ ] Fall from height	– Non-ionizing
[ ] Struck by moving/falling object	[ ] Drowning
	[ ] Motor vehicle accident
[ ] Caught in/between object	[ ] Others
[ ] Electrical contact	

8. Use the information from the steps above to identify and list the hazards (Figure 7.6) and describe each in relation to each task.

9. Use the information from other sections of the manual together with the information collected above to identify the root causes of the hazards. Figure 7.3 illustrates the process whereby information collected is analysed to help identify the root causes of the hazards to assist in developing solutions.

119

7.5 Make design decisions

The objective of this step is to develop design solutions in the safety working group, assess risks, define technical and cost implications, and select the most appropriate solution based on the balance between the seriousness of the hazard and the implications of the solutions.

1. List all hazards from steps 7.3 and 7.4, with their likely root causes.
2. List possible design solutions and evaluate them for costs and technical consequences. Use the following sequence of priorities, in order, to develop and evaluate the solution:
   (a) Design to eliminate hazard.
   (b) Design to minimize hazard.
   (c) Provide barriers.
   (d) Provide warning devices.
   (e) Provide formal work methods.
3. Select design solution (a) which eliminates the hazard, if there are no significant implications on the design.
4. If there are problems/significant implications, pursue a risk assessment. One procedure to prioritize risks is given below. It can be modified to suit local circumstances but definitions and criteria should be agreed by the safety working group and written into the group's procedure, for later reference and review.
   (a) Use Table 7.1 to assess the potential severity of injury, H, M or L.
   (b) Use Table 7.2 to obtain a value for the frequency of exposure to the hazard, H, M or L.
   (c) Using these two values, use Table 7.3 to find the hazard level rating, 1, 2 or 3.

Table 7.1 Severity of potential hazards.

Severity rating	Definition
High (H)	– Likely to cause death, permanent total disability or long term injury or health problem. – High task demands and high risk of musculoskeletal injury. Lost time injury is likely to most workers.
Medium (M)	– Death is not likely but still a consideration. – Permanent partial disability is possible. – Task requirements would exceed the mental and physical capabilities of some workers. Lost time injury is likely to these workers.
Low (L)	– Death or major injury is very unlikely. Minor injuries possible. – Task requirements difficult for some workers but within their capabilities. Lost time injury possible for some workers, but unlikely.

Table 7.2 Frequency of exposure.

Number of people performing the task	Number of times task is performed by each worker or each person is exposed to hazard		
	Daily to weekly	Weekly to bi-monthly	< Bi-monthly
Many	H	H	M
Moderate	H	M	L
Few	M	L	L

Table 7.3 Hazard level rating.

Severity	Frequency of exposure		
	H	M	L
H	Level 3	Level 3	Level 3
M	Level 3	Level 2	Level 2
L	Level 2	Level 1	Level 1

5. Use the hazard level rating to select the appropriate design solution for hazard control:
   Level 3: Design out the hazard.
   Level 2: Design out the hazard or design to minimize the hazard.
   Level 1: Design out or to minimize the hazard; or provide barriers, personal protective equipment, warning devices or, as a last resort set up administrative controls such as formal work procedures.

7.6 Implement solutions and document the decision

Record all decisions and promulgate to designers, design management and users/customers. Specify all hazards which have not been eliminated, and transmit this information to users/customers. Record and assess all subsequent changes. Monitor equipment/system during its lifetime and record all failures.

7.7 Safety standards

Within the European Union (EU), national standards for the safety of machinery (which cover everything with linked parts, with control and power sources, joined together for a specific application) are based on European Directives, and compliance with one set of national standards should ensure acceptance throughout the EU. In the context of equipment design and ergonomics, an important set of regulations is the Supply of Machinery (Safety) Regulations 1992. From 1 January 1996 machinery must comply with these regulations.
Machinery is required to fulfil wide-ranging health and safety requirements, which are listed in the Regulations under forty-five headings. These requirements indicate that relevant ergonomic principles must be utilized to reduce to the minimum possible the hazards, workloads, discomforts, fatigue and psychological stresses of those working with the equipment.
It is a requirement, also, that documented information must be retained by the manufacturer for up to ten years after the manufacture of any equipment, to enable a technical file to be compiled. This must include, amongst other items, a list of the essential health and safety requirements taken into account in the design, a description of the methods adopted to eliminate hazards and a copy of the instructions for the machinery. The manufacturer must pursue the necessary research or tests to demonstrate that the machine can be erected, installed, maintained and operated safely.

The requirement for an instruction manual is also mandatory, and it must be in a form which can be readily understood by those involved in the machine's use. Noise levels or vibration levels created by the machine must be given. In many cases it may be appropriate to incorporate most of the material required for the technical file into the instruction manual. These regulations apply to all machinery, but there are certain specific requirements for especially dangerous equipment, e.g. woodworking machinery. The declaration that the machine conforms to the regulations can be provided by the manufacturer except for this special class of machines, for which an example must be inspected and passed by an 'approved body', as described in the Regulations.

Bibliography

Bell, C.R., 1974, *Men at Work*, London: George Allen & Unwin.

Booth, B., 1967, Technical considerations in selecting handwheels for machine tools, *MTIRA Research Report No. 27*.

BS EN 292 (Parts 1 and 2), 1992, *Basic Concepts and General Principles for Safety in the Design of Machinery*, London: British Standards Institution.

BS EN 294, 1992, *Safety of Machinery: Safety Distances to Prevent Danger Zones being Reached by the Upper Limbs*, London: British Standards Institution.

BS 350, Part 1, 1974, *Conversion Factors and Tables: Basis of Tables*, London: British Standards Institution.

BS 1376, 1985, *Specifications for Colours of Light Signals*, London: British Standards Institution.

BS 2961, 1967, *Typeface Nomenclature and Classification*, London: British Standards Institution.

BS 3042, 1992, *Test Probes to Verify Protection by Enclosures*, London: British Standards Institution.

BS 3641 (Part 1), 1990, *Symbols for Machine Tools*, London: British Standards Institution.

BS 3641 (Part 2), 1987, *Specification for Numerical Control Symbols*, London: British Standards Institution.

BS 3641 (Part 3), 1990, *Additional Symbols for Machine Tools*, London: British Standards Institution.

BS 3641 Supplement, 1987, *Guide to Application of Additional Symbols*, and *Index*, London: British Standards Institution.

BS 3693, 1992, *Recommendations for Design of Scales and Indexes on Analogue Indicating Instruments*, London: British Standards Institution.

BS 4099 (Parts I and II), 1986, *Specification for Colours of Indicator Lights, Push Buttons, Flashing Lights, Annunciators and Digital Readouts*, London: British Standards Institution.

BS 4884 (Part I), 1992, *Specification for Technical Manuals*, London: British Standards Institution.

BS 4884 (Part II), 1983, *Specification for Technical Manuals*, London: British Standards Institution.

BS 5775, 1987, *Specification for Quantities, Units and Symbols*, London: British Standards Institution.

BS 5775 (Part 0), 1987, (ISO 31.0, 1981) *Specification for Quantities, Units and Symbols: General Principles*, London: British Standards Institution.

BS 6805, 1987, (ISO 7574, 1985) *Statistical Methods for Determining and Verifying Stated Noise Emission Values of Machinery and Equipment*, London: British Standards Institution.

BS 6841, 1987, *Guide to Measurement and Evaluation of Human Exposure to Whole Body Mechanical Vibration and Repeated Shock*, London: British Standards Institution.

BS 6842, 1987, *Guide to Measurement and Evaluation of Human Exposure to Vibration Transmitted to the Hand*, London: British Standards Institution.

BS 7324, 1990, (ISO 7000, 1989) *Guide to Graphical Symbols for Use on Equipment*, London: British Standards Institution.

BS 7477, 1991, *Guide for General Principles for the Creation of Graphic Symbols for Use on Equipment*, London: British Standards Institution.

BS EN 60204, 1992, *Safety of Machinery: Electrical Equipment of Industrial Machines*, London: British Standards Institution.

BS EN 60204, 1992, *Safety of Machinery: Electrical Equipment of Machine Tools MA4: Valve Handwheels*, London: British Standards Institution.

BS 7445 (Parts 1, 2 and 3), 1991, *Description and Measurement of Environmental Noise*, London: British Standards Institution.

BS 7643, 1993, (ISO 6242.1, 1992) *Building Construction, Specification of Users' Requirements. Part 1, Thermal Requirements*, London: British Standards Institution.

BSI, annual, *British Standards Yearbook*, London: British Standards Institution.

Burns, W., 1973, *Noise and Man*, London: John Murray.

Cakir, A., Hart, D.J. and Stewart, T.F.M., 1981, *Visual Display Terminals – A Manual Covering Ergonomics, Workplace Design, Health and Safety*, New York: J. Wiley & Sons.

CIBS, 1984, *Code for Interior Lighting*, London: Chartered Institution of Building Services.

Damon, A., Stoudt, H.W. and McFarland, R.A., 1966, *The Human Body in Equipment Design*, Cambridge, Mass.: Harvard University Press.

DD 202, 1991, *Ergonomic Principles in the Design of Work Systems*, London: British Standards Institution.

Diffrient, Tilley, Bardagjy, Henry Dreyfuss Associates, 1974, *Humanscale 1/2/3*, Cambridge, Mass.: MIT Press.

DIN 31001, Part I, 1983, *Safety Distances for Adults and Children*, Berlin: Deutches Institut für Normung.

Easterby, R.S., 1970, The perception of symbols for machine displays, *Ergonomics*, **13**, 149–158.

EN 60073, 1993, *Specification for Coding of Indicating Devices and Actuators by Colours and Supplementary Means*, London: British Standards Institution.

Fanger, P.O., 1970, *Thermal Comfort: Analysis and Applications in Environmental Engineering*, Copenhagen: Danish Technical Press.

Galer, I.A.R., 1987, *Applied Ergonomics Handbook*, Second Edition, London: Butterworths.

Health and Safety Executive, 1987, *Lighting at Work*, HS(G)38, London: HMSO.

Health and Safety Executive, 1989, *Human Factors in Industial Safety*, HS(G)48, London: HMSO.

Health and Safety Executive, 1990, *Work-related Upper Limb Disorders: A Guide to Prevention*, London: HMSO.

Health and Safety Executive, 1991, *Seating at Work*, HS(G)57, London: HMSO.

Health and Safety Executive, 1992, *Work Equipment: Guidance on Regulations*, L.22, London: HMSO.

Health and Safety Executive, 1992, *Manual Handling: Guidance on Regulations*, L.23, London: HMSO.

Her Majesty's Factory Inspectorate, 1971, Noise and the Worker, *Health and Safety at Work Reports No. 25*, London: HMSO.

Her Majesty's Factory Inspectorate, 1978, Lighting in Offices, Shops and Railway Premises, *Health and Safety at Work Reports No. 39*, London: HMSO.

Hopkinson, R.G. and Collins, J.B., 1977, *The Ergonomics of Lighting*, London: Macdonald Technical Scientific.

ISO 5349, 1986, Guidelines for the Measurement and Assessment of Human Exposure to Hand Transmitted Vibration, Geneva: International Standards Organization.

ISO 7730, 1984, *Moderate Thermal Environments – Determination of the PMV and PPD Indices and Specification of the Conditions for Thermal Comfort*, Geneva: International Standards Organization.

ISO, annual, *International Standards Organization Directory*, Geneva: International Standards Organization.

ISO 1996, Part 1 (1982) and Parts 2 and 3 (1987), Geneva: International Standards Organization.

Jones, J.C., 1963, Anthropometric data, limitations in use, *Architects Journal Information Library*, 6 Feb., 317–325.

Kirwan, B. and Ainsworth, L.K. (Eds), 1992, *A Guide to Task Analysis*, London: Taylor & Francis Ltd.

Morgan, C.T., Cook, J.S., Chapanis, A. and Lund, M.W., 1963, *Human Engineering Guide to Equipment Design*, New York: McGraw-Hill.

NIOSH, 1981, *Work Practices Guide for Manual Lifting*, Cincinnati, OH.: National Institute of Occupational Safety and Health.

Pheasant, S.T., 1995, *Bodyspace – Anthropometry, Ergonomics and Design*, Second Edition, London: Taylor & Francis Ltd.

PP 7310, 1990, *Anthropometrics, an Introduction*, London: British Standards Institution.

Product Standards: Machinery. UK Regulations, 1993, London: Department of Trade and Industry.

Product Standards: Machinery Update, 1993, London: Department of Trade and Industry .

Putz-Anderson, V. (Ed.), 1988, *Cumulative Trauma Disorders: A Manual for Musculoskeletal Diseases of the Upper Limbs*, London: Taylor & Francis Ltd.

Shepherd, A., 1989, Analysis and training in information technology tasks. In: Diaper, D. (Ed.) *Task Analysis for Human–Computer Interaction*, pp. 15–55, Chichester: Ellis Horwood.

Singleton, W.T., 1974, *Man–Machine Systems*, Harmondsworth: Penguin.

Snook, S.H., 1978, The design of manual handling tasks, *Ergonomics*, **21** 963–985.

Stammers, R.B. and Shepherd, A., 1995, Task analysis. In: Wilson, J.R. and Corlett, E.N. (Eds) *Evaluation of Human Work*, second edition, London: Taylor & Francis Ltd.

The Supply of Machinery (Safety) Regulations, 1992, London: HMSO.

Van Cott, H.P., and Kinkade, R.G. (Eds), 1972, *Human Engineering Guide to Equipment Design*, Washington, DC: US Government Printing Office.

Waters, T.R., Putz-Anderson, V., Garg, A. and Fine, L.J., 1993, Revised NIOSH equation for the design evaluation of manual lifting tasks, *Ergonomics*, **36**, 7, 749–776.

Whitfield, D., 1971, British Standards and ergonomics, *Applied Ergonomics*, **2**, 238–242.

Wilson, J.R. and Corlett, E.N. (Eds), 1995, *Evaluation of Human Work*, second edition, London: Taylor & Francis Ltd.

Wright, P., 1971, Writing to be understood: why use sentences? *Applied Ergonomics*, **2**, 207–242.

Wright, P. and Fox, K., 1970, Presenting information in tables, *Applied Ergonomics*, **1**, 232–242.

Index